Patterns, Predictions, and Actions

Patterns, Predictions, and Actions

Foundations of Machine Learning

Moritz Hardt and Benjamin Recht

Princeton University Press
Princeton and Oxford

Published by Princeton University Press, 41 William Street,
Princeton, New Jersey 08540

In the United Kingdom: Princeton University Press, 99 Banbury Road, Oxford OX2 6JX

press.princeton.edu

ISBN 9780691233734
ISBN (e-book) 9780691233727

British Library Cataloging-in-Publication Data is available

Editorial: Hallie Stebbins, Kiran Pandey
Production Editorial: Terri O'Prey
Cover Design: Wanda España
Production: Jacqueline Poirier
Publicity: Matthew Taylor, Charlotte Coyne
Cover image by ElenaBs / Alamy Stock Vector

This book has been composed in LaTeX.

The publisher would like to acknowledge the authors of this volume for providing the print-ready files from which this book was printed.

Printed on acid-free paper. ∞

Printed in the United States of America

1 3 5 7 9 10 8 6 4 2

For Isaac, Leonora, and Quentin

Contents

List of Figures

List of Tables

Preface

In its conception, our book is both an old take on something new and a new take on something old.

Looking at it one way, we return to the roots with our emphasis on pattern classification. We believe that the practice of machine learning today is surprisingly similar to pattern classification of the 1960s, with a few notable innovations from more recent decades.

This is not to understate recent progress. Like many, we are amazed by the advances that have happened in recent years. Image recognition has improved dramatically. Even small devices can now reliably recognize speech. Natural language processing and machine translation have made massive leaps forward. Machine learning has even been helpful in some difficult scientific problems, such as protein folding.

However, we think that it would be a mistake not to recognize pattern classification as a driving force behind these improvements. The ingenuity behind many advances in machine learning so far lies not in a fundamental departure from pattern classification, but rather in finding new ways to make problems amenable to the model fitting techniques of pattern classification.

Consequently, the first few chapters of this book follow relatively closely the excellent text *Pattern Classification and Scene Analysis* by Duda and Hart, particularly, its first edition from 1973, which remains relevant today. Indeed, Duda and Hart summarize the state of pattern classification in 1973, and it bears a striking resemblance to the core of what we consider today to be machine learning. We add new developments on representations, optimization, and generalization, all of which remain topics of evolving, active research.

Looking at it differently, our book departs in some considerable ways from the way machine learning is commonly taught.

First, our text emphasizes the role that datasets play in machine learning. A full chapter explores the histories, significance, and scientific basis of machine learning benchmarks. Although ubiquitous and taken for granted today, the datasets-as-benchmarks paradigm was a relatively recent development of the 1980s. Detailed consideration of datasets, the collection and construction of data, as well as the training and testing paradigm, tend to be lacking from theoretical courses on machine learning.

Second, the book includes a modern introduction to causality and the practice of causal inference that lays to rest dated controversies in the field. The introduction is self-contained, starts from first principles, and requires no prior commitment intellectually or ideologically to the field of causality. Our treatment of causality includes the conceptual foundations, as well as some of the practical tools of causal inference increasingly applied in numerous applications. It's interesting to note that many recent causal estimators reduce the problem of causal inference in clever ways to pattern classification. Hence, this material fits quite well with the rest of the book.

Third, our book covers sequential and dynamic models thoroughly. Though such material could easily fill a semester course on its own, we wanted to provide the basic elements required to think about making decisions in dynamic contexts. In particular, given so much recent interest in reinforcement learning, we hope to provide a self-contained short introduction to the concepts underpinning this field. Our approach here follows our approach to supervised learning: we focus on how we would make decisions given a probabilistic model of our environment, and then turn to how to take action when the model is unknown. Hence, we begin with a focus on optimal sequential decision making and dynamic programming. We describe some of the basic solution approaches to such problems, and discuss some of the complications that arise as our measurement quality deteriorates. We then turn to making decisions when our models are unknown, providing a survey of bandit optimization and reinforcement learning. Our focus here is to again highlight the power of prediction. We show that for most problems, pattern recognition can be seen as a complement to feedback control, and we highlight how "certainty equivalent" decision making—where we first use data to estimate a model and then use feedback control acting as if this model were true—is optimal or near-optimal in a surprising number of scenarios.

Finally, we attempt to highlight in a few different places throughout the potential harms, limitations, and social consequences of machine learning. From its roots in World War II, machine learning has always been political. Advances in artificial intelligence feed into a global industrial military complex, and are funded by it. As useful as machine learning is for some unequivocally positive applications such as assistive devices, it is also used to great effect for tracking, surveillance, and warfare. Commercially its most successful use cases to date are targeted advertising and digital content recommendation, both of questionable value to society. Several scholars have explained how the use of machine learning can perpetuate inequity through the ways that it can put additional burden on already marginalized, oppressed, and disadvantaged communities. Narratives of artificial intelligence also shape policy in several high-stakes debates about the replacement of human judgment in favor of statistical models in the criminal justice system, health care, education, and social services.

There are some notable topics we left out. Some might find that the most glaring omission is the lack of material on unsupervised learning. Indeed, there has been a significant amount of work on unsupervised learning in recent years. Thankfully, some of the most successful approaches to learning without labels could be described as *reductions to pattern recognition*. For example, researchers have found ingenious ways of procuring labels from unlabeled data points, an approach called self supervision. We believe that the contents of this book will prepare students interested in these topics well.

The material we cover supports a one semester graduate introduction to machine learning. We invite readers from all backgrounds. However, mathematical maturity with probability, calculus, and linear algebra is required. We provide a chapter on mathematical background for review. Necessarily, this chapter cannot replace prerequesite coursework.

In writing this book, our goal was to balance mathematical rigor and insights we have found useful in the most direct way possible. In contemporary learning theory, important results often have short sketches, yet making these arguments rigorous and precise may require dozens of pages of technical calculations. Such proofs are critical to the community's scientific activities but often make important insights hard to access for those not yet versed in the appropriate techniques. On the other hand, many machine learning courses drop proofs altogether, thereby losing the important foundational ideas that they contain. We aim to strike a balance, including full details for as many arguments as possible, but frequently referring readers to the relevant literature for full details.

Acknowledgments

We are indebted to Alexander Rakhlin, who pointed us to the early generalization bound for the Perceptron algorithm. This result both in its substance and historical position shaped our understanding of machine learning. Kevin Jamieson was the first to point out to us the similarity between the structure of our course and the text by Duda and Hart. Peter Bartlett provided many helpful pointers to the literature and historical context of generalization theory. Jordan Ellenberg helped us improve the presentation of algorithmic stability. Dimitri Bertsekas pointed us to an elegant proof of the Neyman-Pearson Lemma. We are grateful to Rediet Abebe and Ludwig Schmidt for discussions relating to the chapter on datasets. We also are grateful to David Aha, Thomas Dietterich, Michael I. Jordan, Pat Langley, John Platt, and Csaba Szepesvari for giving us additional context about the state of machine learning in the 1980s. Finally, we are indebted to Boaz Barak, David Blei, Adam Klivans, Csaba Szepesvari, and Chris Wiggins for detailed feedback and suggestions on an early draft of this text. We're also grateful to Chris Wiggins for pointing us to Highleyman's data.

We thank all students of UC Berkeley's CS 281a in the Fall of 2019, 2020, and 2021, who worked through various iterations of the material in this book. Special thanks to our graduate student instructors Mihaela Curmei, Sarah Dean, Frances Ding, Sara Fridovich-Keil, Wenshuo Guo, Chloe Hsu, Meena Jagadeesan, John Miller, Robert Netzorg, Juan C. Perdomo, and Vickie Ye, who spotted and corrected many mistakes we made.

We are grateful for comments and feedback from Namhoon Cho, Yanai Elazar, Thomas Fork, Awni Hannun, Michaela Hardt, Nadia Hyder, Toru Lin, Yuxi Liu, Diego Marez, Thomas Tian, and Ramon Vilarino.

Patterns, Predictions, and Actions

Chapter 1

Introduction

"Reflections on life and death of those who in Breslau lived and died" is the title of a manuscript that Protestant pastor Caspar Neumann sent to mathematician Gottfried Wilhelm Leibniz in the late seventeenth century. Neumann had spent years keeping track of births and deaths in his Polish hometown, now called Wrocław. Unlike sprawling cities like London and Paris, Breslau had a rather small and stable population with limited migration in and out. The parishes in town took due record of the newly born and deceased.

Neumann's goal was to find patterns in the occurrence of births and deaths. He thereby sought to dispel a persisting superstition that ascribed critical importance to certain climacteric years of age. Some believed it was at age 63, others that it was either at the 49th or the 81st year, that particularly critical events threatened to end the journey of life. Neumann recognized that his data defied the existence of such climacteric years.

Leibniz must have informed the Royal Society of Neumann's work. In turn, the Society invited Neumann in 1691 to provide the Society with the data he had collected. It was through the Royal Society that British astronomer Edmund Halley became aware of Neumann's work. A friend of Isaac Newton's, Halley had spent years predicting the trajectories of celestial bodies, but not those of human lives.

After a few weeks of processing the raw data through smoothing and interpolation, it was in the spring of 1693 that Halley arrived at what became known as Halley's life table.

At the outset, Halley's table displayed, for each year of age, the number of people of that age alive in Breslau at the time. Halley estimated that a total of approximately 34,000 people were alive, of which approximately 1,000 were between the ages zero and one, 855 were between ages one and two, and so forth.

Halley saw multiple applications of his table. One of them was to estimate the proportion of men in a population that could bear arms. To estimate this

Age. Curt.	Persons.	Age. Curt.	Persons	Age. Curt.	Persons	Age. Curt.	Persons	Age. Curt.	Persons	Age. Curt.	Persons
1	1000	8	680	15	628	22	585	29	539	36	481
2	855	9	670	16	622	23	579	30	531	37	472
3	798	10	661	17	616	24	573	31	523	38	463
4	760	11	653	18	610	25	567	32	515	39	454
5	732	12	646	19	604	26	560	33	507	40	445
6	710	13	640	20	598	27	553	34	499	41	436
7	692	14	634	21	592	28	546	35	490	42	427
Age. Curt	Persons.	Age. Curt.	Persons	Age. Curt.	Persons	Age Curt	Persons	Age. Curt.	Persons	Age. Curt.	Persons
43	417	50	346	57	272	64	202	71	131	78	58
44	407	51	335	58	262	65	192	72	120	79	49
45	397	52	324	59	252	66	182	73	109	80	41
46	387	53	313	60	242	67	172	74	98	81	34
47	377	54	302	61	232	68	162	75	88	82	28
48	367	55	292	62	222	69	152	76	78	83	23
49	357	56	282	63	212	70	142	77	68	84	20

Age.	Persons.
7	5547
14	4584
21	4270
28	3964
35	3604
42	3178
49	2709
56	2194
63	1694
70	1204
77	692
84	253
100	107
	34000
	Sum Total.

Figure 1.1: Halley's life table

proportion he computed the number of people between age 18 and 56, and divided by 2. The result suggested that 26% of the population were men neither too old nor too young to go to war.

At the same time, King William III of England needed to raise money for his country's continued involvement in the Nine Years War, raging from 1688 to 1697. In 1692, William turned to a financial innovation imported from Holland, the public sale of life annuities. A life annuity is a financial product that pays out a predetermined annual amount of money while the purchaser of the annuity is alive. The king had offered annuities at fourteen times the annual payout, a price too low for the young and too high for the old.

Halley recognized that his table could be used to estimate the odds that a person of a certain age would die within the next year. Based on this observation, he described a formula for pricing an annuity that, expressed in modern language, computes the sum of expected discounted payouts over the course of a person's life starting from their current age.

Ambitions of the twentieth century

Halley had stumbled upon the fact that prediction requires no physics. Unknown outcomes, be they future or unobserved, often follow patterns found in past observations. This empirical law would become the basis of consequential decision making for centuries to come.

On the heels of Halley and his contemporaries, the eighteenth century saw the steady growth of the life insurance industry. The industrial revolution fueled other forms of insurance sold to a population seeking safety in tumultuous times. Corporations and governments developed risk models of increasing complexity with varying degrees of rigor. Actuarial science and financial risk assessment became major fields of study built on the empirical law.

Modern statistics and decision theory emerged in the late nineteenth and early twentieth century. Statisticians recognized that the scope of the empirical law extended far beyond insurance pricing, that it could be a method for both scientific discovery and decision making writ large.

Emboldened by advances in probability theory, statisticians modeled populations as probability distributions. Attention turned to what a scientist could say about a population by looking at a random draw from its probability distribution. From this perspective, it made sense to study how to decide between one of two plausible probability models for a population in light of available data. The resulting concepts, such as true positive and false positive, as well as the resulting technical repertoire, are in broad use today as the basis of hypothesis testing and binary classification.

As statistics flourished, two other developments around the middle of the twentieth century turned out to be transformational. The works of Turing, Gödel, and von Neumann, alongside dramatic improvements in hardware, marked the beginning of the computing revolution. Computer science emerged as a scientific discipline. General-purpose programmable computers promised a new era of automation with untold possibilities.

World War II spending fueled massive research and development programs on radar, electronics, and servomechanisms. Established in 1940, the United States National Defense Research Committee included a division devoted to control systems. The division developed a broad range of control systems, including gun directors, target predictors, and radar-controlled devices. The agency also funded theoretical work by mathematician Norbert Wiener, including plans for an ambitious anti-aircraft missile system that used statistical methods for predicting the motion of enemy aircraft.

In 1948, Wiener published his influential book *Cybernetics* around the same time as Shannon published *A Mathematical Theory of Communication*. Both proposed theories of information and communication, but their goals were different. Wiener's ambition was to create a new science, called cybernetics, that unified communications and control in one conceptual framework. Wiener believed that there was a close analogy between the human nervous system and digital computers. He argued that the principles of control, communication, and feedback could be a way not only to create mind-like machines, but also to understand the interaction of machines and humans. Wiener even went so far as to posit that the dynamics of entire social systems and civilizations could be understood and steered through the organizing principles of cybernetics.

The zeitgeist that animated cybernetics also drove ambitions to create artificial neural networks, capable of carrying out basic cognitive tasks. Cognitive concepts such as learning and intelligence had entered research conversations about computing machines and with them came the quest for machines that learn from experience.

The 1940s were a decade of active research on artificial neural networks, often called connectionism. A 1943 paper by McCulloch and Pitts formalized artificial neurons and provided theoretical results about the universality of artificial neural networks as computing devices. A 1949 book by Donald Hebb pursued the central idea that neural networks might learn by constructing internal representations of concepts.

Pattern classification

Around the mid 1950s, it seemed that progress on connectionism had started to slow and would have perhaps tapered off had psychologist Frank Rosenblatt not made a striking discovery.

Rosenblatt had devised a machine for image classification. Equipped with 400 photosensors, the machine could read an image composed of 20 by 20 pixels and sort it into one of two possible classes. Mathematically, the perceptron computes a linear function of its input pixels. If the value of the linear function applied to the input image is positive, the perceptron decides that its input belongs to class 1, otherwise class -1. What made the perceptron so successful was the way it could learn from examples. Whenever it misclassified an image, it would adjust the coefficients of its linear function via a local correction.

Rosenblatt observed in experiments what would soon be a theorem. If a sequence of images could at all be perfectly classified by a linear function, the perceptron would only make so many mistakes on the sequence before it correctly classified all images it encountered.

Rosenblatt developed the perceptron in 1957 and continued to publish on the topic in the years that followed. The perceptron project was funded by the US Office of Naval Research, which jointly announced the project with Rosenblatt at a press conference in 1958 that led the *New York Times* to exclaim:

> The Navy revealed the embryo of an electronic computer that it expects will be able to walk, talk, see, write, reproduce itself and be conscious of its existence.[1]

This development sparked significant interest in perceptrons and reinvigorated neural network research throughout the 1960s. By all accounts, the research in the decade that followed Rosenblatt's work had essentially all the ingredients of what is now called machine learning, specifically, supervised learning.

Practitioners experimented with a range of different features and model architectures, moving from linear functions to perceptrons with multiple layers, the equivalent of today's deep neural networks. A range of variations of the optimization method and different ways of propagating errors came and went.

Theory followed closely behind. Not long after the invention came a theorem, called mistake bound, that gave an upper bound on the number of mistakes the perceptron would make in the worst case on any sequence of labeled data points that can be fit perfectly with a linear separator.

Today, we recognize the perceptron as an instance of the stochastic gradient method applied to a suitable objective function. The stochastic gradient method remains the optimization workhorse of modern machine learning applications.

Shortly after the well-known mistake bound came a lesser known theorem. The result showed that when the perceptron succeeded in fitting training data, it would also succeed in classifying unseen examples correctly provided that these were drawn from the same distribution as the training data. We call this *generalization*: finding rules consistent with available data that apply to instances we have yet to encounter.

By the late 1960s, these ideas from perceptrons had solidified into a broader subject called *pattern recognition* that encompassed most of the concepts we consider core to machine learning today. In 1939, Wald formalized the basic problem of classification as one of optimal decision making when the data is generated by a known probabilistic model. Researchers soon realized that pattern classification could be achieved using data alone to guide prediction methods such as perceptrons, nearest neighbor classifiers, and density estimators. The connections with mathematical optimization including gradient descent and linear programming also took shape during the 1960s.

Pattern classification—today more popularly known as supervised learning—built on statistical tradition in how it formalized the idea of generalization. We assume observations come from a fixed data generating process, such as samples drawn from a fixed distribution. In a first optimization step, called training, we fit a model to a set of data points labeled by class membership. In a second step, called testing, we judge the model by how well it performs on newly generated data from the very same process.

This notion of generalization as performance on fresh data can seem mundane. After all, it simply requires the classifier to do, in a sense, more of the same. We require consistent success on the same data generating process as encountered during training. Yet the seemingly simple question of what theory underwrites the generalization ability of a model has occupied the machine learning research community for decades.

Pattern classification, once again

Machine learning as a field, however, is not a straightforward evolution of the pattern recognition of the 1960s, at least not culturally and not historically.

After a decade of perceptron research, a group of influential researchers, including McCarthy, Minsky, Newell, and Simon, put forward a research program by the name of artificial intelligence. The goal was to create human-like intelligence in a machine. Although the goal itself was in many ways not far from the ambitions of connectionists, the group around McCarthy fancied entirely different formal techniques. Rejecting the numerical pattern fitting of the connectionist era, the proponents of this new discipline saw the future in symbolic and logical manipulation of knowledge represented in formal languages.

Artificial intelligence became the dominant academic discipline to deal with cognitive capacities of machines within the computer science community. Pattern recognition and neural networks research continued, albeit largely outside artificial intelligence. Indeed, journals on pattern recognition flourished during the 1970s.

During this time, artificial intelligence research led to a revolution in *expert systems*, logic- and rule-based models that had significant industrial impact. Expert systems were hard coded and left little room for adapting to new information. AI researchers interested in such adaptation and improvement—learning, if you will—formed their own subcommunity, beginning in 1981 with the first International Workshop on Machine Learning. The early work from this community reflects the logic-based research that dominated artificial intelligence at the time; the papers read as if of a different field than what we now recognize today as machine learning research. It was not until the late 1980s that machine learning began to look more like pattern recognition, once again.

Personal computers had made their way from research labs into home offices across wealthy nations. Internet access, if slow, made email a popular form of communication among researchers. File transfer over the internet allowed researchers to share code and datasets more easily.

Machine learning researchers recognized that in order for the discipline to thrive it needed a way to more rigorously evaluate progress on concrete tasks. Whereas in the 1950s it had seemed miraculous enough if training errors decreased over time on any nontrivial task, it was clear now that machine learning needed better benchmarks.

In the late 1980s, the first widely used benchmarks emerged. Then graduate student David Aha created the UCI machine learning repository that made several datasets widely available via FTP. Aiming to better quantify the performance of AI systems, the Defense Advanced Research Projects Agency (DARPA) funded a research program on speech recognition that led to the creation of the influential TIMIT speech recognition benchmark.

These benchmarks had the data split into two parts, one called training data, one called testing data. This split elicits the promise that the learning algorithm will only access the training data when it fits the model. The testing data is reserved for evaluating the trained model. The research community can then rank learning algorithms by how well the trained models perform on the testing data.

Splitting data into training and testing sets was an old practice, but the idea of reusing such datasets as benchmarks was novel and transformed machine learning. The *dataset-as-benchmark paradigm* caught on and became core to applied machine learning research for decades to come. Indeed, machine learning benchmarks were at the center of the most recent wave of progress on deep learning. Chief among them was ImageNet, a large repository of images, labeled by nouns of objects displayed in the images. A subset of roughly 1 million images belonging to 1,000 different object classes was the basis of the ImageNet Large Scale Visual Recognition Challenge. Organized from 2010 until 2017, the competition became a striking showcase for performance of deep learning methods for image classification.

Increases in computing power and volume of available data were key driving factors for progress in the field. But machine learning benchmarks did more than provide data. Benchmarks gave researchers a way to compare results, share ideas, and organize communities. They implicitly specified a problem description and a minimal interface contract for code. Benchmarks also became a means of knowledge transfer between industry and academia.

The most recent wave of machine learning as pattern classification was so successful, in fact, that it became the new artificial intelligence in the public narrative of popular media. The technology reached entirely new levels of commercial significance with companies competing fiercely over advances in the space.

This new artificial intelligence had done away with the symbolic reasoning of the McCarthy era. Instead, the central drivers of progress were widely regarded as growing datasets, increasing compute resources, and more benchmarks along with publicly available code to start from. Are those then the only ingredients needed to secure the sustained success of machine learning in the real world?

Prediction and action

Unknown outcomes often follow patterns found in past observations. But what do we do with the patterns we find and the predictions we make? Like Halley proposing his life table for annuity pricing, predictions only become useful when they are acted upon. But going from patterns and predictions to successful actions is a delicate task. How can we even anticipate the effect of a hypothetical action when our actions now influence the data we observe and value we accrue in the future?

One way to determine the effect of an action is experimentation: try it out and see what happens. But there's a lot more we can do if we can model the situation more carefully. A model of the environment specifies how an action changes the state of the world, and how in turn this state results in a gain or loss of utility. We include some aspects of the environment explicitly as variables in our model. Others we declare *exogenous* and model as noise in our system.

The solution of how to take such models and turn them into plans of action that maximize expected utility is a mathematical achievement of the twentieth century. By and large, such problems can be solved by *dynamic programming*. Initially formulated by Bellman in 1954, dynamic programming poses optimization problems where at every time step, we observe data, take an action, and pay a cost. By chaining these steps together in time, elaborate plans can be made that remain optimal under considerable stochastic uncertainty. These ideas revolutionized aerospace in the 1960s, and are still deployed in infrastructure planning, supply chain management, and the landing of SpaceX rockets. Dynamic programming remains one of the most important algorithmic building blocks in the computer science toolkit.

Planning actions under uncertainty has also always been core to artificial intelligence research, though initial proposals for sequential decision making in AI were more inspired by neuroscience than operations research. In 1950-era AI, the main motivating concept was one of *reinforcement learning*, which posited that one should encourage taking actions that were successful in the past. This reinforcement strategy led to impressive game-playing algorithms like Samuel's Checkers Agent circa 1959. Surprisingly, it wasn't until the 1990s that researchers realized that reinforcement learning methods were approximation schemes for dynamic programming. Powered by this connection, a mix of researchers from AI and operations research applied neural nets and function approximation to simplify the approximate solution of dynamic programming problems. The subsequent 30 years have led to impressive advances in reinforcement learning and approximate dynamic programming techniques for playing games, such as Go, and in powering dexterous manipulation in robotic systems.

Central to the reinforcement learning paradigm is understanding how to balance learning about an environment and acting on it. This balance is a nontrivial problem even in the case where actions do not lead to a change in state. In the context of machine learning, experimentation in the form of taking an action and observing its effect often goes by the name *exploration*. Exploration reveals the payoff of an action, but it comes at the expense of not taking an action that we already knew had a decent payoff. Thus, there is an inherent trade-off between exploration and *exploitation* of previous actions. Though in theory the optimal balance can be computed by dynamic programming, it is more common to employ techniques from *bandit optimization* that are simple and effective strategies to balance exploration and exploitation.

Not limited to experimentation, causality is a comprehensive conceptual framework to reason about the effect of actions. Causal inference, in principle, allows us to estimate the effect of hypothetical actions from observational data. A growing technical repertoire of causal inference is taking various sciences by storm, as witnessed in epidemiology, political science, policy, climate, and development economics.

There are good reasons that many see causality as a promising avenue for making machine learning methods more robust and reliable. Current state-of-the-art predictive models remain surprisingly fragile to changes in the data. Even small natural variations in a data-generating process can significantly deteriorate performance. There is hope that tools from causality could lead to machine learning methods that perform better under changing conditions.

However, causal inference is no panacea. There are no causal insights without making substantive judgments about the problem that are not verifiable from data alone. The reliance on hard earned substantive domain knowledge stands in contrast with the nature of recent advances in machine learning that largely did without it—and that was the point.

Chapter notes

Halley's life table has been studied and discussed extensively; for an entry point, see recent articles by Bellhouse[2] and Ciecka,[3] or the article by Pearson and Pearson.[4]

Halley was not the first to create a life table. In fact, what Halley created is more accurately called a population table. Instead, John Grount deserves credit for the first life table in 1662 based on mortality records from London. Considered to be the founder of demography and an early epidemiologist, Grount's work was in many ways more detailed than Halley's fleeting engagement with Breslau's population. However, to Grount's disadvantage the mortality records released in London at the time did not include the age of the deceased, thus complicating the work significantly.

Mathematician de Moivre picked up Halley's life table in 1725 and sharpened the mathematical rigor of Halley's idea. A few years earlier, de Moivre had published the first textbook on probability theory called *The Doctrine of Chances: A Method of Calculating the Probability of Events in Play*. Although de Moivre lacked the notion of a probability distribution, his book introduced an expression resembling the normal distribution as an approximation to the binomial distribution, what was in effect the first central limit theorem. The time of Halley coincides with the emergence of probability. Hacking's book provides much additional context, particularly relevant are Chapter 12 and 13.[5]

For the history of feedback, control, and computing before cybernetics, see the excellent text by Mindell.[6] For more on the cybernetics era itself, see the books by Kline[7] and Heims.[8] See Beniger[9] for how the concepts of control and communication and the technology from that era led to the modern information society.

The prologue from the 1988 edition of *Perceptrons* by Minsky and Papert presents a helpful historical perspective. The recent 2017 reprint of the same book contains additional context and commentary in a foreword by Léon Bottou.

Much of the first International Workshop on Machine Learning was compiled in an edited volume, which summarizes the motivations and perspectives that seeded the field.[10] Langley's article provides helpful context on the state of evaluation in machine learning in the 1980s and how the desire for better metrics led to a renewed emphasis on pattern recognition.[11] Similar calls for better evaluation motivated the speech transcription program at DARPA, leading to the TIMIT dataset, arguably the first machine learning benchmark dataset.[12,13,14]

It is worth noting that the Parallel Distributed Processing Research Group led by Rummelhart and McLeland actively worked on neural networks during the 1980s and made extensive use of the rediscovered back-propagation algorithm, an efficient algorithm for computing partial derivatives of a circuit.[15]

A recent article by Jordan provides an insightful perspective on how the field came about and what challenges it still faces.[16]

Chapter 2

Fundamentals of Prediction

Prediction is the art and science of leveraging patterns found in natural and social processes to conjecture about uncertain events. We use the word *prediction* broadly to refer to statements about things we don't know for sure *yet*, including but not limited to the outcome of future events.

Machine learning is to a large extent the study of algorithmic prediction. Before we can dive into machine learning, we should familiarize ourselves with prediction. Starting from first principles, we will motivate the goals of prediction before building up to a statistical theory of prediction.

We can formalize the goal of prediction problems by assuming a population of N instances with a variety of attributes. We associate with each instance two variables, denoted X and Y. The goal of prediction is to conjecture a plausible value for Y after observing X alone. But when is a prediction good? For that, we must quantify some notion of the quality of prediction and aim to optimize that quantity.

To start, suppose that for each variable X we make a deterministic prediction $f(X)$ by means of some prediction function f. A natural goal is to find a function f that makes the fewest number of incorrect predictions, where $f(X) \neq Y$, across the population. We can think of this function as a computer program that reads X as input and outputs a prediction $f(X)$ that we hope matches the value Y. For a fixed prediction function, f, we can sum up all of the errors made on the population. Dividing by the size of the population, we observe the average (or mean) error rate of the function.

Minimizing errors

Let's understand how we can find a prediction function that makes as few errors as possible on a given population in the case of binary prediction, where the variable Y has only two values.

Consider a population of Abalone, a type of marine snail with colorful shells featuring a varying number of rings. Our goal is to predict the sex,

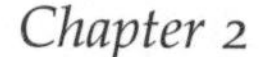

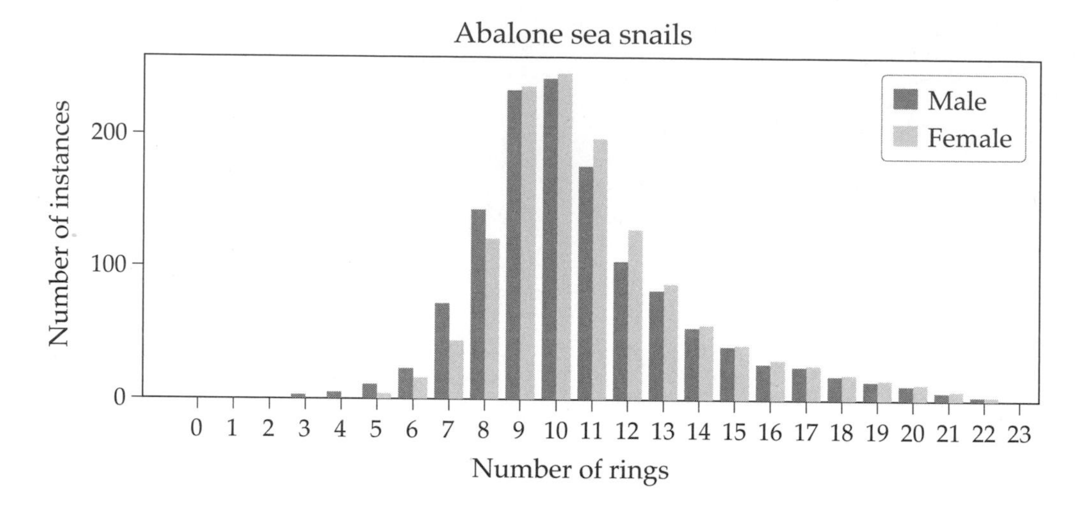

Figure 2.1: Histogram of Abalone sea snail population

male or female, of the Abalone from the number of rings on the shell. We can tabulate the population of Abalone by counting for each possible number of rings the number of male and female instances in the population.

From this way of presenting the population, it is not hard to compute the predictor that makes the fewest mistakes. For each value on the X-axis, we predict "female" if the number of female instances with this X-value is larger than the number of male instances. Otherwise, we predict "male" for the given X-value. For example, there's a majority of male Abalone with seven rings on the shell. Hence, it makes sense to predict "male" when we see seven rings on a shell. Scrutinizing the figure a bit further, we can see that the best possible predictor is a *threshold function* that returns "male" whenever the number of rings is at most 8, and "female" whenever the number of rings is greater or equal to 9.

The number of mistakes our predictor makes is still significant. After all, most counts are pretty close to each other. But it's better than random guessing. It uses whatever there is that we can say from the number of rings about the sex of an Abalone snail, which is just not much.

What we constructed here is called the *minimum error rule*. It generalizes to multiple attributes. If we had measured not only the number of rings, but also the length of the shell, we would repeat the analogous counting exercise over the two-dimensional space of all possible values of the two attributes.

The minimum error rule is intuitive and simple, but computing the rule exactly requires examining the entire population. Tracking down every instance of a population is not only intractable. It also defeats the purpose of prediction in almost any practical scenario. If we had a way of enumerating the X and Y values of all instances in a population, the prediction problem would be solved.

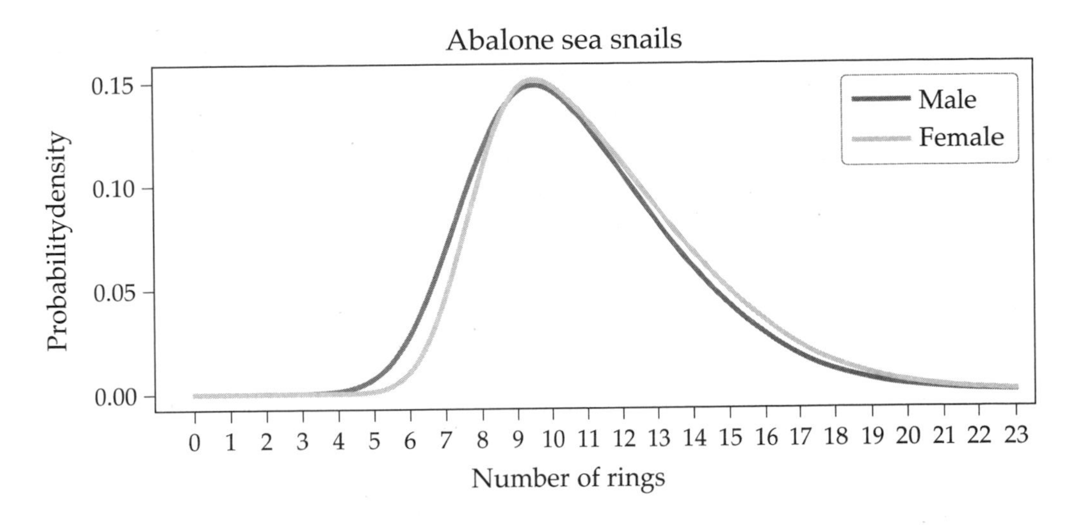

Figure 2.2: Representing Abalone population as a distribution

Given an instance X we could simply look up the corresponding value of Y from our records.

What's missing so far is a way of doing prediction that does not require us to enumerate the entire population of interest.

Modeling knowledge

Fundamentally, what makes prediction without enumeration possible is *knowledge* about the population. Human beings organize and represent knowledge in different ways. In this chapter, we will explore in depth the consequences of one particular way to represent populations, specifically, as *probability distributions*.

The assumption we make is that we have knowledge of a probability distribution $p(x, y)$ over pairs of X and Y values. We assume that this distribution conceptualizes the "typical instance" in a population. If we were to select an instance uniformly at random from the population, what relations between its attributes might we expect? We expect that a uniform sample from our population would be the same as a sample from $p(x, y)$. We call such a distribution a *statistical model* or simply *model* of a population. The word *model* emphasizes that the distribution isn't the population itself. It is, in a sense, a sketch of a population that we use to make predictions.

Let's revisit our Abalone example in probabilistic form. Assume we know the distribution of the number of rings of male and female Abalone, as illustrated in the figure.

Both follow a skewed normal distribution described by three parameters each, a location, a scale, and a skew parameter. Knowing the distribution is to assume that we know these parameters. Although the specific numbers won't matter for our example, let's spell them out for concreteness. The distribution for male Abalone has location 7.4, scale 4.48, and skew 3.12, whereas the distribution for female Abalone has location 7.63, scale 4.67, and skew 4.34. To complete the specification of the joint distribution over X and Y, we need to determine the relative proportion of males and females. Assume for this example that male and female Abalone are equally likely.

Representing the population this way, it makes sense to predict "male" whenever the probability density for male Abalone is larger than that for female Abalone. By inspecting the plot we can see that the density is higher for male snails up until 8 rings, at which point it is larger for female instances. We can see that the predictor we derive from this representation is the same threshold rule that we had before.

We arrived at the same result without the need to enumerate and count all possible instances in the population. Instead, we recovered the minimum error rule from knowing only seven parameters, three for each conditional distribution, and one for the balance of the two classes.

Modeling populations as probability distributions is an important step in making prediction algorithmic. It allows us to represent populations succinctly, and gives us the means to make predictions about instances we haven't encountered.

Subsequent chapters extend these fundamentals of prediction to the case where we don't know the exact probability distribution, but only have a random sample drawn from the distribution. It is tempting to think about machine learning as being all about *that*, namely what we do with a sample of data drawn from a distribution. However, as we learn in this chapter, many fundamentally important questions arise even if we have full knowledge of the population.

Prediction from statistical models

Let's proceed to formalize prediction assuming we have full knowledge of a statistical model of the population. Our first goal is to formally develop the minimum error rule in greater generality.

We begin with binary prediction where we suppose Y has two alternative values, 0 and 1. Given some measured information X, our goal is to conjecture whether Y equals 0 or 1.

Throughout we assume that X and Y are random variables drawn from a joint probability distribution. It is convenient both mathematically and conceptually to specify the joint distribution as follows. We assume that Y has a priori (or *prior*) probabilities:

$$p_0 = \mathbb{P}[Y = 0], \qquad p_1 = \mathbb{P}[Y = 1].$$

That is, we assume we know the proportion of instances with $Y = 1$ and $Y = 0$ in the population. We'll always model available information as being a random

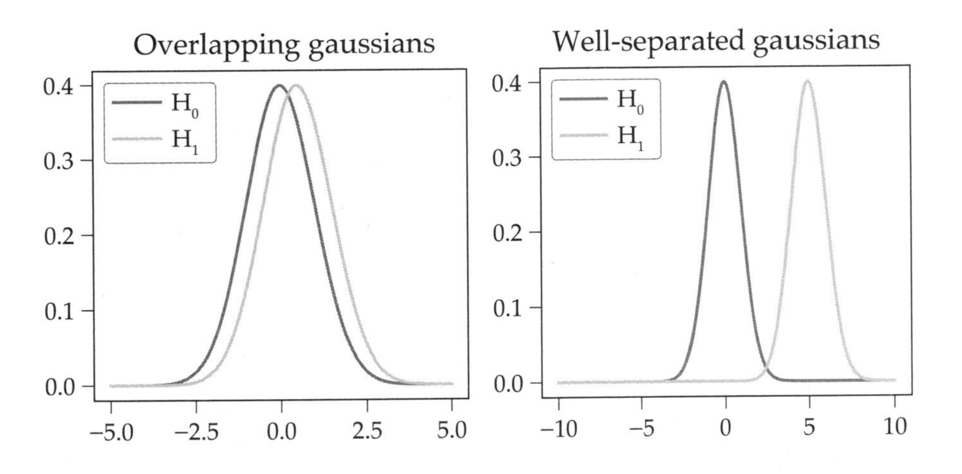

Figure 2.3: Illustration of shifted Gaussians

vector X with support in $\mathbb{R}^d$. Its distribution depends on whether Y is equal to 0 or 1. In other words, there are two different statistical models for the data, one for each value of Y. These models are the conditional probability densities of X given a value y for Y, denoted $p(x \mid Y = y)$. This density function is often called a *generative model* or *likelihood function* for each scenario.

Example: Signal versus noise

For a simple example with more mathematical formalism, suppose that when $Y = 0$ we observe a scalar $X = \omega$ where ω is unit-variance, zero-mean Gaussian noise $\omega \sim \mathcal{N}(0, 1)$. Recall that the Gaussian distribution of mean μ and variance σ^2 is given by the density $\frac{1}{\sigma\sqrt{2\pi}} e^{-\frac{1}{2}\left(\frac{x-\mu}{\sigma}\right)^2}$.

Suppose when $Y = 1$, we would observe $X = s + \omega$ for some scalar s. That is, the conditional densities are

$$p(x \mid Y = 0) = \mathcal{N}(0, 1),$$
$$p(x \mid Y = 1) = \mathcal{N}(s, 1).$$

The larger the shift s is, the easier it is to predict whether $Y = 0$ or $Y = 1$. For example, suppose $s = 10$ and we observed $X = 11$. If we had $Y = 0$, the probability that the observation is greater than 10 is on the order of 10^{-23}, and hence we'd likely think we're in the alternative scenario where $Y = 1$. However, if s were very close to zero, distinguishing between the two alternatives is rather challenging. We can think of a small difference s that we're trying to detect as a *needle in a haystack*.

Prediction via optimization

Our core approach to all statistical decision making will be to formulate an appropriate optimization problem for which the decision rule is the optimal solution. That is, we will optimize over *algorithms*, searching for functions that map data to decisions and predictions. We will define an appropriate notion of the cost associated to each decision, and attempt to construct decision rules that minimize the expected value of this cost. As we will see, choosing this optimization framework has many immediate consequences.

Predictors and labels

A *predictor* is a function $\widehat{Y}(x)$ that maps an input x to a prediction $\widehat{y} = \widehat{Y}(x)$. The prediction $\widehat{y}$ is also called a *label* for the point x. The target variable Y can be both real valued or discrete. When Y is a discrete random variable, each different value it can take on is called a *class* of the prediction problem.

To ease notation, we take the liberty to write $\widehat{Y}$ as a shorthand for the random variable $\widehat{Y}(X)$ that we get by applying the prediction function $\widehat{Y}$ to the random variable X.

The most common case we consider through the book is binary prediction, where we have two classes, 0 and 1. Sometimes it's mathematically convenient to instead work with the numbers -1 and 1 for the two classes.

In most cases we consider, labels are scalars that are either discrete or real valued. Sometimes it also makes sense to consider vector-valued predictions and target variables.

The creation and encoding of suitable labels for a prediction problem is an important step in applying machine learning to real-world problems. We will return to it multiple times.

Loss functions and risk

The final ingredient in our formal setup is a *loss function* that generalizes the notion of an error, which we defined as a mismatch between prediction and target values.

A *loss function* takes two inputs, $\widehat{y}$ and y, and returns a real number $loss(\widehat{y}, y)$ that we interpret as a quantified loss for predicting $\widehat{y}$ when the target is y. A loss could be negative, in which case we think of it as a reward.

A prediction error corresponds to the loss function $loss(\widehat{y}, y) = \mathbb{1}\{\widehat{y} \neq y\}$ that indicates disagreement between its two inputs. Loss functions give us modeling flexibility that will become crucial as we apply this formal setup throughout this book.

An important notion is the expected loss of a predictor taken over a population. This construct is called *risk*.

Definition 1. *We define the* risk *associated with* $\hat{Y}$ *to be*

$$R[\hat{Y}] := \mathbb{E}[loss(\hat{Y}(X), Y)] .$$

Here, the expectation is taken jointly over X and Y.

Now that we have defined risk, our goal is to determine which predictor minimizes risk. Let's get a sense for how we might go about this.

In order to minimize risk, theoretically speaking, we need to solve an *infinite dimensional* optimization problem over binary-valued functions. That is, *for every x*, we need to find a binary assignment. Fortunately, the infinite dimension here turns out to not be a problem analytically once we make use of the law of iterated expectation.

Lemma 1. *The optimal predictor is given by*

$$\hat{Y}(x) = \mathbb{1}\left\{ \mathbb{P}[Y = 1 \mid X = x] \geq \frac{loss(1,0) - loss(0,0)}{loss(0,1) - loss(1,1)} \, \mathbb{P}[Y = 0 \mid X = x] \right\} .$$

This rule corresponds to the intuitive rule we derived when thinking about how to make predictions over the population. For a fixed value of the data $X = x$, we compare the frequency that $Y = 1$ occurs to the frequency that $Y = 0$ occurs. If the ratio of these frequencies exceeds some threshold defined by our loss function, then we set $\hat{Y}(x) = 1$. Otherwise, we set $\hat{Y}(x) = 0$.

Proof. To see why this is rule is optimal, we make use of the law of iterated expectation:

$$\mathbb{E}[loss(\hat{Y}(X), Y)] = \mathbb{E}\left[\mathbb{E}\left[loss(\hat{Y}(X), Y) \mid X \right] \right] .$$

Here, the outer expectation is over a random draw of X and the inner expectation samples Y conditional on X. Since there are no constraints on the predictor $\hat{Y}$, we can minimize the expression by minimizing the inner expectation independently for each possible setting that X can assume.

Indeed, for a fixed value x, we can expand the expected loss for each of the two possible predictions:

$$\mathbb{E}[loss(0, Y) \mid X = x] = loss(0,0)\, \mathbb{P}[Y = 0 \mid X = x] + loss(0,1)\, \mathbb{P}[Y = 1 \mid X = x]$$
$$\mathbb{E}[loss(1, Y) \mid X = x] = loss(1,0)\, \mathbb{P}[Y = 0 \mid X = x] + loss(1,1)\, \mathbb{P}[Y = 1 \mid X = x] .$$

The optimal assignment for this x is to set $\hat{Y}(x) = 1$ whenever the second expression is smaller than the first. Writing out this inequality and rearranging gives us the rule specified in the lemma.

□

Probabilities of the form $\mathbb{P}[Y = y \mid X = x]$, as they appeared in the lemma, are called *posterior* probability.

We can relate them to the likelihood function via Bayes' rule:

$$\mathbb{P}[Y = y \mid X = x] = \frac{p(x \mid Y = y)p_y}{p(x)},$$

where $p(x)$ is a density function for the marginal distribution of X.

When we use posterior probabilities, we can rewrite the optimal predictor as

$$\widehat{Y}(x) = \mathbb{1}\left\{\frac{p(x \mid Y = 1)}{p(x \mid Y = 0)} \geq \frac{p_0(\mathit{loss}(1,0) - \mathit{loss}(0,0))}{p_1(\mathit{loss}(0,1) - \mathit{loss}(1,1))}\right\}.$$

This rule is an example of a likelihood ratio test.

Definition 2. *The* likelihood ratio *is the ratio of the likelihood functions:*

$$\mathcal{L}(x) := \frac{p(x \mid Y = 1)}{p(x \mid Y = 0)}.$$

A likelihood ratio test *(LRT) is a predictor of the form*

$$\widehat{Y}(x) = \mathbb{1}\{\mathcal{L}(x) \geq \eta\}$$

for some scalar threshold $\eta > 0$.

If we denote the optimal threshold value

$$\eta = \frac{p_0(\mathit{loss}(1,0) - \mathit{loss}(0,0))}{p_1(\mathit{loss}(0,1) - \mathit{loss}(1,1))}, \qquad (1)$$

then the predictor that minimizes the risk is the likelihood ratio test

$$\widehat{Y}(x) = \mathbb{1}\{\mathcal{L}(x) \geq \eta\}.$$

An LRT naturally partitions the sample space in two regions:

$$\mathcal{X}_0 = \{x \in \mathcal{X} : \mathcal{L}(x) \leq \eta\},$$
$$\mathcal{X}_1 = \{x \in \mathcal{X} : \mathcal{L}(x) > \eta\}.$$

The sample space $\mathcal{X}$ then becomes the disjoint union of $\mathcal{X}_0$ and $\mathcal{X}_1$. Since we only need to identify which set x belongs to, we can use any function $h : \mathcal{X} \to \mathbb{R}$ that gives rise to the same threshold rule. As long as $h(x) \leq t$ whenever $\mathcal{L}(x) \leq \eta$ and vice versa, these functions give rise to the same partition into $\mathcal{X}_0$ and $\mathcal{X}_1$. So, for example, if g is any monotonically increasing function, then the predictor

$$\widehat{Y}_g(x) = \mathbb{1}\{g(\mathcal{L}(x)) \geq g(\eta)\}$$

is equivalent to using $\hat{Y}(x)$. In particular, it's popular to use the logarithmic predictor

$$\hat{Y}_{\log}(x) = \mathbb{1}\{\log p(x \mid Y=1) - \log p(x \mid Y=0) \geq \log(\eta)\},$$

as it is often more convenient or numerically stable to work with logarithms of likelihoods.

This discussion shows that there are an *infinite number of functions* that give rise to the same binary predictor. Hence, we don't need to know the conditional densities exactly and can still compute the optimal predictor. For example, suppose the true partitioning of the real line under an LRT is

$$\mathcal{X}_0 = \{x \colon x \geq 0\} \quad \text{and} \quad \mathcal{X}_1 = \{x \colon x < 0\}.$$

Setting the threshold to $t = 0$, the functions $h(x) = x$ or $h(x) = x^3$ give the same predictor, as does any odd function that is positive on the right half line.

Example: Needle in a haystack revisited

Let's return to our needle in a haystack example with

$$p(X \mid Y=0) = \mathcal{N}(0,1),$$
$$p(X \mid Y=1) = \mathcal{N}(s,1),$$

and assume that the prior probability of $Y=1$ is very small, say, $p_1 = 10^{-6}$. Suppose that if we declare $\hat{Y}=0$, we do not pay a cost. If we declare $\hat{Y}=1$ but are wrong, we incur a cost of 100. But if we guess $\hat{Y}=1$ and it is actually true that $Y=1$, we actually gain a reward of 1,000,000. That is, $loss(0,0) = 0$, $loss(0,1) = 0$, $loss(1,0) = 100$, and $loss(1,1) = -1{,}000{,}000$.

What is the LRT for this problem? Here, it's considerably easier to work with logarithms:

$$\log(\eta) = \log\left(\frac{(1-10^{-6}) \cdot 100}{10^{-6} \cdot 10^{6}}\right) \approx 4.61.$$

Now,

$$\log p(x \mid Y=1) - \log p(x \mid Y=0) = -\frac{1}{2}(x-s)^2 + \frac{1}{2}x^2 = sx - \frac{1}{2}s^2.$$

Hence, the optimal predictor is to declare

$$\hat{Y} = \mathbb{1}\left\{sx > \tfrac{1}{2}s^2 + \log(\eta)\right\}.$$

The optimal rule here is *linear*. Moreover, the rule divides the space into two open intervals. While the entire real line lies in the union of these two intervals, it is exceptionally unlikely to ever see an x larger than $|s| + 5$. Hence, even if our predictor were incorrect in these regions, the risk would still be nearly optimal as these terms have almost no bearing on our expected risk!

Maximum a posteriori and maximum likelihood

A folk theorem of statistical decision theory states that essentially all optimal rules are equivalent to likelihood ratio tests. While this isn't *always* true, many important prediction rules end up being equivalent to LRTs. Shortly, we'll see an optimization problem that speaks to the power of LRTs. But before that, we can already show that the well-known *maximum likelihood* and *maximum a posteriori* predictors are both LRTs.

The expected error of a predictor is the expected number of times we declare $\widehat{Y} = 0$ (resp. $\widehat{Y} = 1$) when $\widehat{Y} = 1$ (resp. $\widehat{Y} = 0$) is true. Minimizing the error is equivalent to minimizing the risk with cost $\mathit{loss}(0,0) = \mathit{loss}(1,1) = 0$, $\mathit{loss}(1,0) = \mathit{loss}(0,1) = 1$. The optimum predictor is hence a likelihood ratio test. In particular,

$$\widehat{Y}(x) = \mathbb{1}\left\{\mathcal{L}(x) \geq \tfrac{p_0}{p_1}\right\}.$$

Using Bayes' rule, one can see that this rule is equivalent to

$$\widehat{Y}(x) = \arg \max_{y \in \{0,1\}} \mathbb{P}[Y = y \mid X = x].$$

Recall that the expression $\mathbb{P}[Y = y \mid X = x]$ is called the posterior probability of $Y = y$ given $X = x$. And this rule is hence referred to as the *maximum a posteriori* (MAP) rule.

As we discussed above, the expression $p(x \mid Y = y)$ is called the *likelihood* of the point x given the class $Y = y$. A maximum likelihood rule would set

$$\widehat{Y}(x) = \arg\max_y p(x \mid Y = y).$$

This is completely equivalent to the LRT when $p_0 = p_1$ and the costs are $\mathit{loss}(0,0) = \mathit{loss}(1,1) = 0$, $\mathit{loss}(1,0) = \mathit{loss}(0,1) = 1$. Hence, the maximum likelihood rule is equivalent to the MAP rule with a uniform prior on the labels.

That both of these popular rules ended up reducing to LRTs is no accident. In what follows, we will show that LRTs are almost always the optimal solution of optimization-driven decision theory.

Types of errors and successes

Let $\widehat{Y}(x)$ denote any predictor mapping into $\{0,1\}$. Binary predictions can be right or wrong in four different ways, summarized by the *confusion table*.

Table 2.1: Confusion table

	$Y = 0$	$Y = 1$
$\hat{Y} = 0$	true negative	false negative
$\hat{Y} = 1$	false positive	true positive

Taking expected values over the populations gives us four corresponding *rates* that are characteristics of a predictor.

1. **True Positive Rate:** $\text{TPR} = \mathbb{P}[\hat{Y}(X) = 1 \mid Y = 1]$. Also known as *power, sensitivity, probability of detection*, or *recall*.
2. **False Negative Rate:** $\text{FNR} = 1 - \text{TPR}$. Also known as *type II error* or *probability of missed detection*.
3. **False Positive Rate:** $\text{FPR} = \mathbb{P}[\hat{Y}(X) = 1 \mid Y = 0]$. Also known as *size* or *type I error* or *probability of false alarm*.
4. **True Negative Rate** $\text{TNR} = 1 - \text{FPR}$, the probability of declaring $\hat{Y} = 0$ given $Y = 0$. This is also known as *specificity*.

There are other quantities that are also of interest in statistics and machine learning:

1. **Precision:** $P[Y = 1 \mid \hat{Y}(X) = 1]$. This is equal to $(p_1\text{TPR})/(p_0\text{FPR} + p_1\text{TPR})$.
2. **F1-score:** F_1 is the harmonic mean of precision and recall. We can write this as
$$F_1 = \frac{2\text{TPR}}{1 + \text{TPR} + \frac{p_0}{p_1}\text{FPR}}\,.$$
3. **False discovery rate:** False discovery rate (FDR) is equal to the expected ratio of the number of false positives to the total number of positives.

In the case where both labels are equally likely, precision, F_1, and FDR are also only functions of FPR and TPR. However, these quantities explicitly account for *class imbalances*: when there is a significant skew between p_0 and p_1, such measures are often preferred.

TPR and FPR are competing objectives. We'd like TPR as large as possible and FPR as small as possible.

We can think of risk minimization as optimizing a balance between TPR and FPR:

$$R[\hat{Y}] := \mathbb{E}[loss(\hat{Y}(X), Y)] = \alpha\text{FPR} - \beta\text{TPR} + \gamma\,,$$

where α and β are nonnegative and γ is some constant. For all such α, β, and γ, the risk-minimizing predictor is an LRT.

Other cost functions might try to balance TPR versus FPR in other ways. Which pairs of (FPR, TPR) are achievable?

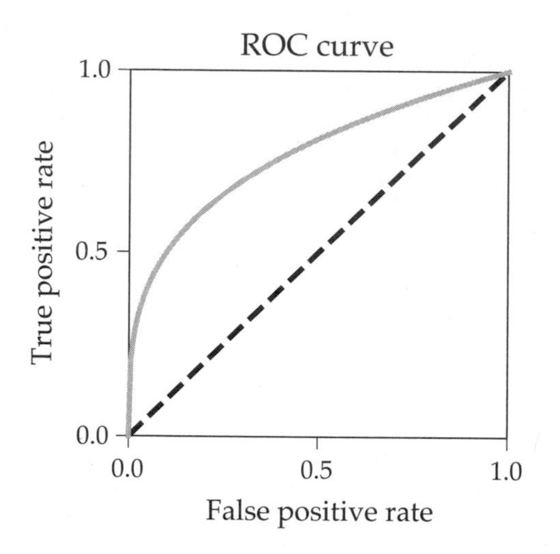

Figure 2.4: Example of an ROC curve

ROC curves

True and false positive rates lead to another fundamental notion, called the the *receiver operating characteristic (ROC) curve.*

The ROC curve is a property of the joint distribution (X, Y) and shows for every possible value $\alpha = [0, 1]$ the best possible true positive rate that we can hope to achieve with any predictor that has false positive rate α. As a result the ROC curve is a curve in the FPR-TPR plane. It traces out the maximal TPR for any given FPR. Clearly the ROC curve contains values $(0, 0)$ and $(1, 1)$, which are achieved by constant predictors that either reject or accept all inputs.

We will now show, in a celebrated result by Neyman and Pearson, that the ROC curve is given by varying the threshold in the likelihood ratio test from negative to positive infinity.

The Neyman-Pearson Lemma

The Neyman-Pearson Lemma, a fundamental lemma of decision theory, will be an important tool for us to establish three important facts. First, it will be a useful tool for understanding the geometric properties of ROC curves. Second, it will demonstrate another important instance where an optimal predictor is a likelihood ratio test. Third, it introduces the notion of probabilistic predictors.

Suppose we want to maximize the true positive rate subject to an upper bound on the false positive rate. That is, we aim to solve the optimization problem:

$$\begin{array}{ll} \text{maximize} & \text{TPR} \\ \text{subject to} & \text{FPR} \leq \alpha\,. \end{array}$$

Let's optimize over *probabilistic predictors*. A probabilistic predictor Q returns 1 with probability $Q(x)$ and 0 with probability $1 - Q(x)$. With such rules, we can rewrite our optimization problem as:

$$\begin{array}{ll} \text{maximize}_Q & \mathbb{E}[Q(X) \mid Y = 1] \\ \text{subject to} & \mathbb{E}[Q(X) \mid Y = 0] \leq \alpha \\ & \forall x \colon Q(x) \in [0,1] \end{array}$$

Lemma 2. ***Neyman-Pearson Lemma.*** *Suppose the likelihood functions $p(x|y)$ are continuous. Then the optimal probabilistic predictor that maximizes* TPR *with an upper bound on* FPR *is a deterministic likelihood ratio test.*

Even in this constrained setup, allowing for more powerful probabilistic rules, we can't escape likelihood ratio tests. The Neyman-Pearson Lemma has many interesting consequences in its own right that we will discuss momentarily. But first, let's see why the lemma is true.

The key insight is that for any LRT, we can find a loss function for which it is optimal. We will prove the lemma by constructing such a problem, and using the associated condition of optimality.

Proof. Let η be the threshold for an LRT such that the predictor

$$Q_\eta(x) = \mathbb{1}\{\mathcal{L}(x) > \eta\}$$

has FPR $= \alpha$. Such an LRT exists because we assumed our likelihoods were continuous. Let β denote the TPR of Q_η.

We claim that Q_η is optimal for the risk minimization problem corresponding to the loss function

$$loss(1,0) = \tfrac{\eta p_1}{p_0},\ loss(0,1) = 1,\ loss(1,1) = 0,\ loss(0,0) = 0\,.$$

Indeed, recalling Equation 1, the risk minimizer for this loss function corresponds to a likelihood ratio test with threshold value

$$\frac{p_0(loss(1,0) - loss(0,0))}{p_1(loss(0,1) - loss(1,1))} = \frac{p_0 loss(1,0)}{p_1 loss(0,1)} = \eta\,.$$

Moreover, under this loss function, the risk of a predictor Q equals

$$\begin{aligned} R[Q] &= p_0 \mathrm{FPR}(Q) loss(1,0) + p_1(1 - \mathrm{TPR}(Q)) loss(0,1) \\ &= p_1 \eta \mathrm{FPR}(Q) + p_1(1 - \mathrm{TPR}(Q))\,. \end{aligned}$$

Now let Q be any other predictor with $\mathrm{FPR}(Q) \leq \alpha$. We have by the optimality of Q_η that

$$\begin{aligned} p_1 \eta \alpha + p_1(1 - \beta) &\leq p_1 \eta \mathrm{FPR}(Q) + p_1(1 - \mathrm{TPR}(Q)) \\ &\leq p_1 \eta \alpha + p_1(1 - \mathrm{TPR}(Q))\,, \end{aligned}$$

which implies $\mathrm{TPR}(Q) \leq \beta$. This in turn means that Q_η maximizes TPR for all rules with $\mathrm{FPR} \leq \alpha$, proving the lemma.

□

Properties of ROC curves

A specific randomized predictor that is useful for analysis combines two other rules. Suppose the first predictor yields $(\mathrm{FPR}^{(1)}, \mathrm{TPR}^{(1)})$ and the second predictor achieves $(\mathrm{FPR}^{(2)}, \mathrm{TPR}^{(2)})$. If we flip a biased coin and use the first rule with probability p and the second rule with probability $1-p$, then this yields a randomized predictor with $(\mathrm{FPR}, \mathrm{TPR}) = (p\mathrm{FPR}^{(1)} + (1-p)\mathrm{FPR}^{(2)}, p\mathrm{TPR}^{(1)} + (1-p)\mathrm{TPR}^{(2)})$. Using this rule lets us prove several properties of ROC curves.

Proposition 1. *The points $(0,0)$ and $(1,1)$ are on the ROC curve.*

Proof. This proposition follows because the point $(0,0)$ is achieved when the threshold $\eta = \infty$ in the likelihood ratio test, corresponding to the constant 0 predictor. The point $(1,1)$ is achieved when $\eta = 0$, corresponding to the constant 1 predictor. □

The Neyman-Pearson Lemma gives us a few more useful properties.

Proposition 2. *The ROC must lie above the main diagonal.*

Proof. To see why this proposition is true, fix some $\alpha > 0$. Using a randomized rule, we can achieve a predictor with $\mathrm{TPR} = \mathrm{FPR} = \alpha$. But the Neyman-Pearson LRT with FPR constrained to be less than or equal to α achieves true positive rate greater than or equal to the randomized rule. □

Proposition 3. *The ROC curve is concave.*

Proof. Suppose $(\mathrm{FPR}(\eta_1), \mathrm{TPR}(\eta_1))$ and $(\mathrm{FPR}(\eta_2), \mathrm{TPR}(\eta_2))$ are achievable. Then

$$(t\mathrm{FPR}(\eta_1) + (1-t)\mathrm{FPR}(\eta_2), t\mathrm{TPR}(\eta_1) + (1-t)\mathrm{TPR}(\eta_2))$$

is achievable by a randomized test. Fixing $\mathrm{FPR} \le t\mathrm{FPR}(\eta_1) + (1-t)\mathrm{FPR}(\eta_2)$, we see that the optimal Neyman-Pearson LRT achieves $\mathrm{TPR} \ge \mathrm{TPR}(\eta_1) + (1-t)\mathrm{TPR}(\eta_2)$. □

Example: The needle one more time

Consider again the *needle in a haystack* example, where $p(x \mid Y=0) = \mathcal{N}(0, \sigma^2)$ and $p(x \mid Y=1) = \mathcal{N}(s, \sigma^2)$ with s a positive scalar. The optimal predictor is to declare $\widehat{Y} = 1$ when X is greater than $\gamma := \frac{s}{2} + \frac{\sigma^2 \log \eta}{s}$. Hence we have

$$\mathrm{TPR} = \int_\gamma^\infty p(x \mid Y=1)\,\mathrm{d}x = \tfrac{1}{2}\operatorname{erfc}\left(\frac{\gamma - s}{\sqrt{2}\sigma}\right)$$
$$\mathrm{FPR} = \int_\gamma^\infty p(x \mid Y=0)\,\mathrm{d}x = \tfrac{1}{2}\operatorname{erfc}\left(\frac{\gamma}{\sqrt{2}\sigma}\right).$$

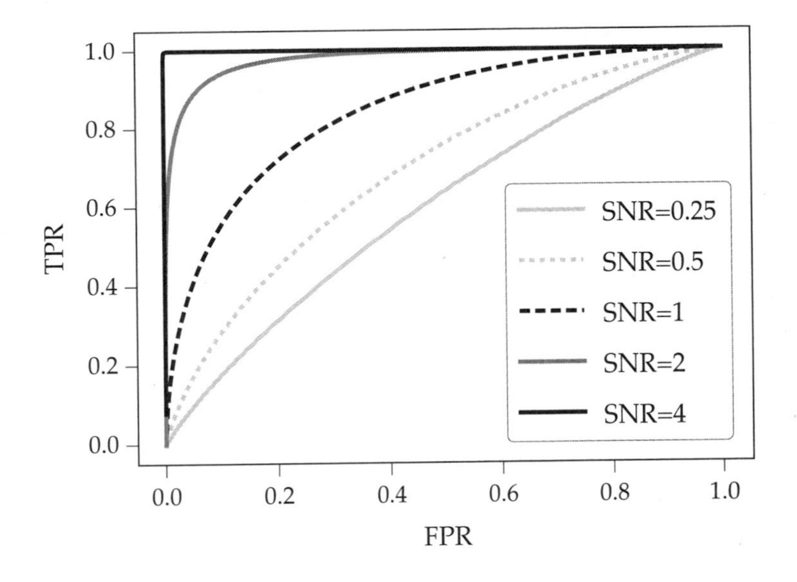

Figure 2.5: ROC curves for various signal-to-noise ratios in the needle in the haystack problem

For fixed s and σ, the ROC curve $(\text{FPR}(\gamma), \text{TPR}(\gamma))$ only depends on the *signal-to-noise ratio* (SNR), s/σ. For small SNR, the ROC curve is close to the FPR = TPR line. For large SNR, TPR approaches 1 for all values of FPR.

Area under the ROC curve

Oftentimes in information retrieval and machine learning, the term ROC curve is overloaded to describe the achievable FPR-TPR pairs that we get by varying the threshold t in any predictor $\hat{Y}(x) = \mathbb{1}\{R(x) > t\}$. Note such curves must lie below the ROC curves that are traced out by the optimal likelihood ratio test, but may approximate the true ROC curves in many cases.

A popular summary statistic for evaluating the quality of a decision function is the area under its associated ROC curve. This is commonly abbreviated as AUC. In the ROC curve plotted in the previous section, as the SNR increases, the AUC increases. However, AUC does not tell the entire story. Here we plot two ROC curves with the same AUC.

If we constrain FPR to be less than 10%, for the light-dotted curve, TPR can be as high as 80% whereas it can only reach 50% for the dark dash-dotted curve. AUC should be always viewed skeptically: the shape of an ROC curve is always more informative than any individual number.

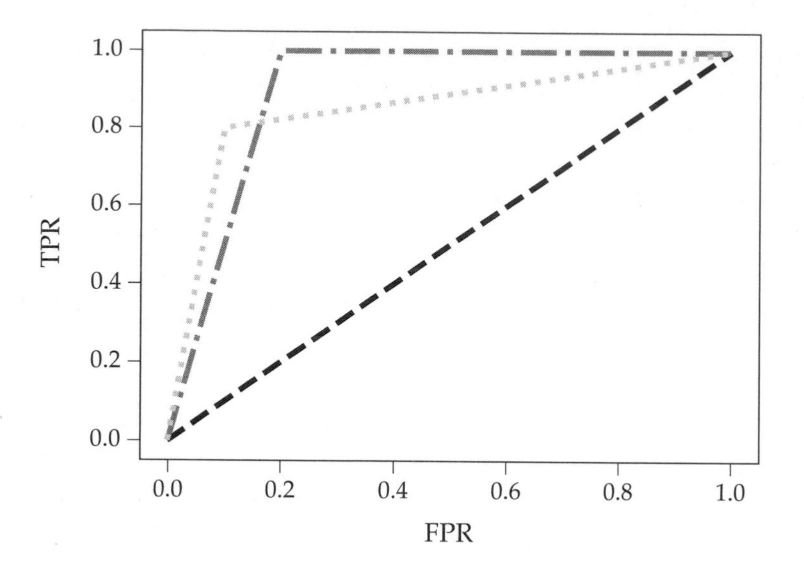

Figure 2.6: Two ROC curves with the same AUC

Decisions that discriminate

The purpose of prediction is almost always decision making. We build predictors to guide our decision making by acting on our predictions. Many decisions entail a life changing event for the individual. The decision could grant access to a major opportunity, such as college admission, or deny access to a vital resource, such as a social benefit.

Binary decision rules always draw a boundary between one group in the population and its complement. Some are labeled *accept*, others are labeled *reject*. When decisions have serious consequences for the individual, however, this decision boundary is not just a technical artifact. Rather it has moral and legal significance.

The decision maker often has access to data that encode an individual's status in socially salient groups relating to race, ethnicity, gender, religion, disability status, and other categories. These have been used as the basis of adverse treatment, oppression, and denial of opportunity in the past and in many cases to this day.

Some see formal or algorithmic decision making as a neutral mathematical tool. However, numerous scholars have shown how formal models can perpetuate existing inequities and cause harm. In her book on this topic, Ruha Benjamin warns of the

> employment of new technologies that reflect and reproduce existing inequities but that are promoted and perceived as more objective or progressive than the discriminatory systems of a previous era.[17]

The narrow formal setup of this chapter highlights an important and challenging facet of the problems of inequality and injustice. Specifically, we are concerned with decision rules that *discriminate* in the sense of creating an unjustified basis of differentiation between individuals.

A concrete example is helpful. Suppose we want to accept or reject individuals for a job. Suppose we have a perfect estimate of the number of hours an individual is going to work in the next five years. We decide that this a reasonable measure of productivity and so we accept every applicant where this number exceeds a certain threshold. On the face of it, our rule might seem neutral. However, on closer reflection, we realize that this decision rule systematically disadvantages individuals who are more likely than others to make use of their parental leave employment benefit that our hypothetical company offers. We are faced with a conundrum. On the one hand, we trust our estimate of productivity. On the other hand, we consider taking parental leave *morally irrelevant* to the decision we're making. It should not be a disadvantage to the applicant. After all that is precisely the reason why the company is offering a parental leave benefit in the first place.

The simple example shows that statistical accuracy alone is no safeguard against discriminatory decisions. It also shows that ignoring *sensitive attributes* is no safeguard either. So what then is *discrimination* and how can we avoid it? This question has occupied scholars from numerous disciplines for decades. There is no simple answer. Before we go into attempts to formalize discrimination in our statistical decision making setting, it is helpful to take a step back and reflect on what the law says.

Legal background in the United States

The legal frameworks governing decision making differ from country to country, and from one domain to another. We take a glimpse at the situation in the United States, bearing in mind that our description is incomplete and does not transfer to other countries.

Discrimination is not a general concept. It is concerned with socially salient categories that have served as the basis for unjustified and systematically adverse treatment in the past. United States law recognizes certain *protected categories* including race, sex (which extends to sexual orientation), religion, disability status, and place of birth.

Further, discrimination is a domain-specific concept concerned with important opportunities that affect people's lives. Regulated domains include credit (Equal Credit Opportunity Act), education (Civil Rights Act of 1964; Education Amendments of 1972), employment (Civil Rights Act of 1964), housing (Fair Housing Act), and *public accommodation* (Civil Rights Act of 1964). Particularly relevant to machine learning practitioners is the fact that the scope of these regulations extends to marketing and advertising within these domains. An ad for a credit card, for example, allocates access to credit and would therefore fall into the credit domain.

There are different legal frameworks available to a plaintiff that brings forward a case of discrimination. One is called *disparate treatment*, the other is *disparate impact*. Both capture different forms of discrimination. Disparate treatment is about purposeful consideration of group membership with the intention of discrimination. Disparate impact is about unjustified harm, possibly through indirect mechanisms. Whereas disparate treatment is about *procedural fairness*, disparate impact is more about *distributive justice*.

It's worth noting that anti-discrimination law does not reflect one overarching moral theory. Pieces of legislation often came in response to civil rights movements, each hard fought through decades of activism.

Unfortunately, these legal frameworks don't give us a formal definition that we could directly apply. In fact, there is some well-recognized tension between the two doctrines.

Formal non-discrimination criteria

The idea of formal non-discrimination (or *fairness*) criteria goes back to pioneering work of Anne Cleary and other researchers in the educational testing community of the 1960s.[18]

The main idea is to introduce a discrete random variable A that encodes membership status in one or multiple protected classes. Formally, this random variable lives in the same probability space as the other covariates X, the decision $\widehat{Y} = \mathbb{1}\{R > t\}$ in terms of a score R, and the outcome Y. The random variable A might coincide with one of the features in X or correlate strongly with some combination of them.

Broadly speaking, different statistical fairness criteria all equalize some group-dependent statistical quantity across groups defined by the different settings of A. For example, we could ask to equalize acceptance rates across all groups. This corresponds to imposing the constraint for all groups a and b:

$$\mathbb{P}[\widehat{Y} = 1 \mid A = a] = \mathbb{P}[\widehat{Y} = 1 \mid A = b].$$

Researchers have proposed dozens of different criteria, each trying to capture different intuitions about what is *fair*. Simplifying the landscape of fairness criteria, we can say that there are essentially three fundamentally different ones of particular significance:

- Acceptance rate $\mathbb{P}[\widehat{Y} = 1]$
- Error rates $\mathbb{P}[\widehat{Y} = 0 \mid Y = 1]$ and $\mathbb{P}[\widehat{Y} = 1 \mid Y = 0]$
- Outcome frequency given score value $\mathbb{P}[Y = 1 \mid R = r]$

The meaning of the first two as a formal matter is clear given what we already covered. The third criterion needs a bit more motivation. A useful property of score functions is *calibration*, which asserts that $\mathbb{P}[Y = 1 \mid R = r] = r$ for all score values r. In words, we can interpret a score value r as the propensity of positive outcomes among instances assigned the score value r. What the third criterion says is closely related. We ask that the score values have the same meaning in each group. That is, instances labeled r in one group are equally likely to be positive instances as those scored r in any other group.

The three criteria can be generalized and simplified using three different conditional independence statements.

Table 2.2: Non-discrimination criteria

Independence	Separation	Sufficiency
$R \perp A$	$R \perp A \mid Y$	$Y \perp A \mid R$

Each of these applies not only to binary prediction, but also to any set of random variables where the independence statement holds. It's not hard to see that independence implies equality of acceptance rates across groups. Separation implies equality of error rates across groups. And sufficiency implies that all groups have the same rate of positive outcomes given a score value.[19]

Researchers have shown that any two of the three criteria are *mutually exclusive* except in special cases. That means, generally speaking, imposing one criterion forgoes the other two.[20,21]

Although these formal criteria are easy to state and arguably natural in the language of decision theory, their merit as measures of discrimination has been subject of an ongoing debate.

Merits and limitations of a narrow statistical perspective

The tension between these criteria played out in a public debate around the use of risk scores to predict *recidivism* in pretrial detention decisions.

There's a risk score, called COMPAS, used by many jurisdictions in the United States to assess *risk of recidivism* in pretrial bail decisions. Recidivism refers to a person's relapse into criminal behavior. In the United States, a defendant may be either detained or released on bail prior to the trial in court depending on various factors. Judges may detain defendants in part based on this score.

Investigative journalists at ProPublica found that Black defendants face a higher false positive rate, i.e., more Black defendants labeled *high risk* end up not committing a crime upon release than White defendants labeled *high risk*.[22] In other words, the COMPAS score fails the separation criterion.

A company called Northpointe, which sells the proprietary COMPAS risk model, pointed out in return that Black and White defendants have equal recidivism rates *given* a particular score value. That is defendants labeled, say, an '8' for *high risk* would go on to recidivate at a roughly equal rate in either group. Northpointe claimed that this property is desirable so that a judge can interpret scores equally in both groups.[23]

The COMPAS debate illustrates both the merits and limitations of the narrow framing of discrimination as a classification criterion.

On the hand, the error rate disparity gave ProPublica a tangible and concrete way to put pressure on Northpointe. The narrow framing of decision making identifies the decision maker as responsible for their decisions. As such, it can be used to interrogate and possibly intervene in the practices of an entity.

On the other hand, decisions are always part of a broader system that embeds structural patterns of discrimination. For example, a measure of recidivism hinges crucially on existing policing patterns. Crime is only found where policing activity happens. However, the allocation and severity of police force itself has racial bias. Some scholars therefore find an emphasis on statistical criteria rather than structural determinants of discrimination to be limited.

Chapter notes

The theory we covered in this chapter is also called *detection theory* and *decision theory*. Similarly, what we call a predictor throughout has various different names, such as *decision rule* or *classifier*.

The elementary detection theory covered in this chapter has not changed much at all since the 1950s and is essentially considered a "solved problem." Neyman and Pearson invented the likelihood ratio test[24] and later proved their lemma, showing it to be optimal for maximizing true positive rates while controlling false positive rates.[25] Wald followed this work by inventing general Bayesian risk minimization in 1939.[26] Wald's ideas were widely adopted during World War II for the purpose of interpreting RADAR signals, which were often very noisy. Much work was done to improve RADAR operations, and this led to the formalization that the output of a RADAR system (the receiver) should be a likelihood ratio, and a decision should be made based on an LRT. Our proof of Neyman-Pearson's lemma came later, and is due to Bertsekas and Tsitsiklis (Section 9.3 of *Introduction to Probability*[27]).

Our current theory of detection was fully developed by Peterson, Birdsall, and Fox in their report on optimal signal detectability.[28] Peterson, Birdsall, and Fox may have been the first to propose Receiver Operating Characteristics as the means to characterize the performance of a detection system, but these ideas were contemporaneously being applied to better understand psychology and psychophysics as well.[29]

Statistical Signal Detection theory was adopted in the pattern recognition community at a very early stage. Chow proposed using optimal detection theory,[30] and this led to a proposal by Highleyman to approximate the risk by its sample average.[31] This transition from population risk to "empirical" risk gave rise to what we know today as machine learning.

Of course, how decisions and predictions are applied and interpreted remains an active research topic. There is a large amount of literature now on the topic of fairness and machine learning. For a general introduction to the problem and dangers associated with algorithmic decision making not limited to discrimination, see the books by Benjamin,[17] Broussard,[32] Eubanks,[33] Noble,[34] and O'Neil.[35] The technical material in our section on discrimination follows Chapter 2 in the textbook by Barocas, Hardt, and Narayanan.[19]

The Abalone example was derived from data available at the UCI Machine Learning Repository, which we will discuss in more detail in Chapter 8. We modified the data to ease exposition. The actual data does not have an equal number of male and female instances, and the optimal predictor is not exactly a threshold function.

Chapter 3

Supervised Learning

Previously, we talked about the fundamentals of prediction and statistical modeling of populations. Our goal was, broadly speaking, to use available information described by a random variable X to conjecture about an unknown outcome Y.

In the important special case of a binary outcome Y, we saw that we can write an optimal predictor $\widehat{Y}$ as a threshold of some function f:

$$\widehat{Y}(x) = \mathbb{1}\{f(x) > t\}$$

We saw that in many cases the optimal function is a ratio of two likelihood functions.

This optimal predictor has a serious limitation in practice, however. To be able to compute the prediction for a given input, we need to know a probability density function for the positive instances in our problem and also one for the negative instances. But we are often unable to construct or unwilling to assume a particular density function.

As a thought experiment, attempt to imagine what a probability density function over images labeled *cat* might look like. Coming up with such a density function appears to be a formidable task, one that's not intuitively any easier than merely classifying whether an image contains a cat or not.

In this chapter, we transition from a purely mathematical characterization of optimal predictors to an algorithmic framework. This framework has two components. One is the idea of working with finite samples from a population. The other is the theory of supervised learning and it tells us how to use finite samples to build predictors algorithmically.

Sample versus population

Let's take a step back to reflect on the interpretation of the pair of random variables (X, Y) that we've worked with so far. We think of the random variables (X, Y) as modeling a population of instances in our prediction problem. From this pair of random variables, we can derive other random variables such as a predictor $\widehat{Y} = \mathbb{1}\{f(X) > t\}$. All of these are random variables in the same probability space. When we talk about, say, the true positive rate of the predictor $\widehat{Y}$, we make a statement about the joint distribution of (X, Y).

In almost all prediction problems, however, we do not have access to the entire population of instances that we will encounter. Neither do we have a probability model for the joint distribution of the random variables (X, Y). The joint distribution is a theoretical construct that we can reason about, but it doesn't readily tell us what to do when we don't have precise knowledge of the joint distribution.

What knowledge then do we typically have about the underlying population and how can we use it algorithmically to find good predictors? In this chapter we will begin to answer both questions.

First we assume that from past experience we have observed n labeled instances $(x_1, y_1), \ldots, (x_n, y_n)$. We assume that each data point (x_i, y_i) is a draw from the same underlying distribution (X, Y). Moreover, we will often assume that the data points are drawn independently. This pair of assumptions is often called the "i.i.d. assumption," a shorthand for *independent and identically distributed*.

To give an example, consider a population consisting of all currently eligible voters in the United States and some of their features, such as, age, income, state of residence, etc. An *i.i.d. sample* from this population would correspond to a repeated sampling process that selects a uniformly random voter from the entire reference population.

Sampling is a difficult problem with numerous pitfalls that can strongly affect the performance of statistical estimators and the validity of what we learn from data. In the voting example, individuals might be unreachable or decline to respond. Even defining a good population for the problem we're trying to solve is often tricky. Populations can change over time. They may depend on a particular social context, or geography, or may not be neatly characterized by formal criteria. Task yourself with the idea of taking a random sample of spoken sentences in the English language, for example, and you will quickly run into these issues.

In this chapter, as is common in learning theory, we largely ignore these important issues. We instead focus on the significant challenges that remain even if we have a well-defined population and an unbiased sample from it.

Supervised learning

Supervised learning is the prevalent method for constructing predictors from data. The essential idea is very simple. We assume we have labeled data, in this context also called *training examples*, of the form $(x_1, y_1), ..., (x_n, y_n)$, where each *example* is a pair (x_i, y_i) of an *instance* x_i and a corresponding *label* y_i. The notion of *supervision* refers to the availability of these labels.

Given such a collection of labeled data points, supervised learning turns the task of finding a good predictor into an optimization problem involving these data points. This optimization problem is called *empirical risk minimization*.

Recall, in the last chapter we assumed full knowledge of the joint distribution of (X, Y) and analytically found predictors that minimize risk. The risk is equal to the expected value of a *loss function* that quantifies the cost of each possible prediction for a given true outcome. For binary prediction problems, there are four possible pairs of labels, corresponding to true positives, false positives, true negatives, and false negatives. In this case, the loss function boils down to specifying a cost to each of the four possibilities.

More generally, a loss function is a function $loss\colon \mathcal{Y} \times \mathcal{Y} \to \mathbb{R}$, where $\mathcal{Y}$ is the set of values that Y can assume. Whereas previously we focused on the predictor $\widehat{Y}$ as a random variable, in this chapter our focus shifts to the functional form that the predictor has. By convention, we write $\widehat{Y} = f(X)$, where $f\colon \mathcal{X} \to \mathcal{Y}$ is a function that maps from the sample space $\mathcal{X}$ into the label space $\mathcal{Y}$.

Although the random variable $\widehat{Y}$ and the function f are mathematically not the same objects, we will call both a predictor and extend our risk definition to apply the function as well:

$$R[f] = \mathbb{E}\left[loss(f(X), Y)\right] .$$

The main new definition in this chapter is a finite sample analog of the risk, called empirical risk.

Definition 3. *Given a set of labeled data points* $S = ((x_1, y_1), ..., (x_n, y_n))$, *the* empirical risk *of a predictor* $f\colon \mathcal{X} \to \mathcal{Y}$ *with respect to the sample* S *is defined as*

$$R_S[f] = \frac{1}{n} \sum_{i=1}^{n} loss(f(x_i), y_i) .$$

Rather than taking expectation over the population, the empirical risk averages the loss function over a finite sample. Conceptually, we think of the finite sample as something that is in our possession, e.g., stored on our hard disk.

Empirical risk serves as a proxy for the risk. Whereas the risk $R[f]$ is a population quantity—that is, a property of the joint distribution (X, Y) and our predictor f—the empirical risk is a *sample quantity*.

We can think of the empirical risk as the sample average estimator of the risk. When samples are drawn i.i.d., the empirical risk is a random variable that equals the sum of n independent random variables. If losses are bounded, the central limit theorem suggests that the empirical risk approximates the risk for a fixed predictor f.

Regardless of the distribution of S, however, note that we can always compute the empirical risk $R_S[f]$ entirely from the sample S and the predictor f. Since empirical risk a quantity we can compute from samples alone, it makes sense to turn it into an objective function that we can try to minimize numerically.

Empirical risk minimization is the optimization problem of finding a predictor in a given function family that minimizes the empirical risk.

Definition 4. *Given a function class* $\mathcal{F} \subseteq \mathcal{X} \to \mathcal{Y}$, empirical risk minimization *on a set of labeled data points S corresponds to the objective:*

$$\min_{f \in \mathcal{F}} R_S[f] \,.$$

A solution to the optimization problem is called empirical risk minimizer.

There is a tautology relating risk and empirical risk that is good to keep in mind:

$$R[f] = R_S[f] + (R[f] - R_S[f]) \,.$$

Although mathematically trivial, the tautology reveals an important insight. To minimize risk, we can first attempt to minimize empirical risk. If we successfully find a predictor f that achieves small empirical risk $R_S[f]$, we're left worrying about the term $R[f] - R_S[f]$. This term quantifies how much the empirical risk of f underestimates its risk. We call this difference *generalization gap* and it is of fundamental importance to machine learning. Intuitively speaking, it tells us how well the performance of our predictor transfers from seen examples (the training examples) to unseen examples (a fresh example from the population) drawn from the same distribution. This process is called *generalization*.

Generalization is not the only goal of supervised learning. A constant predictor that always outputs 0 generalizes perfectly well, but is almost always entirely useless. What we also need is that the predictor achieves small empirical risk $R_S[f]$. Making the empirical risk small is fundamentally about *optimization*. As a consequence, a large part of supervised learning deals with optimization. For us to be able to talk about optimization, we need to commit to a *representation* of the function class $\mathcal{F}$ that appears in the empirical risk minimization problem. The representation of the function class, as well as the choice of a suitable loss function, determines whether or not we can efficiently find an empirical risk minimizer.

To summarize, introducing empirical risk minimization directly leads to three important questions that we will work through in turn.

- **Representation:** What is the class of functions $\mathcal{F}$ we should choose?

- **Optimization:** How can we efficiently solve the resulting optimization problem?
- **Generalization:** Will the performance of predictor transfer gracefully from seen training examples to unseen instances of our problem?

These three questions are intertwined. Machine learning is not so much about studying these questions in isolation as it is about the often delicate interplay between them. Our choice of representation influences both the difficulty of optimization and our generalization performance. Improvements in optimization may not help, or could even hurt, generalization. Moreover, there are aspects of the problem that don't neatly fall into only one of these categories. The choice of the loss function, for example, affects all of the three questions above.

There are important differences between the three questions. Results in optimization, for example, tend to be independent of the statistical assumptions about the data generating process. We will see a number of different optimization methods that under certain circumstances find either a global or a local minimum of the empirical risk objective. In contrast, to reason about generalization, we need some assumptions about the data generating process. The most common one is the i.i.d. assumption we discussed earlier. We will also see several mathematical frameworks for reasoning about the gap between risk and empirical risk.

Let's start with a foundational example that illustrates these core concepts and their interplay.

A first learning algorithm: The perceptron

As we discussed in the introduction, in 1958 the *New York Times* reported the Office of Naval Research claiming the perceptron algorithm[36] would "be able to walk, talk, see, write, reproduce itself and be conscious of its existence." Let's now dive into this algorithm that seemed to have such unbounded potential.

In introducing this algorithm, let's assume we're in a binary prediction problem with labels in $\{-1, 1\}$ for notational convenience. The perceptron algorithm aims to find a *linear separator* of the data, that is, a hyperplane specified by coefficients $w \in \mathbb{R}^d$ so that all positive examples lie on one side of the hyperplane and all negative ones on the other.

Formally, we can express this as $y_i\langle w, x_i\rangle > 0$. In other words, the linear function $f(x) = \langle w, x\rangle$ agrees in sign with the labels on all training instances (x_i, y_i). In fact, the perceptron algorithm will give us a bit more. Specifically, we require that the sign agreement has some *margin* $y_i\langle w, x_i\rangle \geq 1$. That is, when $y = 1$, the linear function must take on a value of at least 1 and when $y = -1$, the linear function must be at most -1. Once we find such a linear function, our

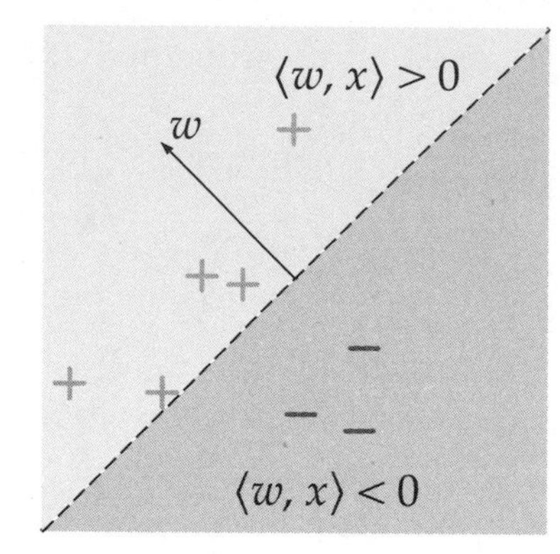

Figure 3.1: Illustration of a linear separator

prediction $\widehat{Y}(x)$ on a data point x is $\widehat{Y}(x) = 1$ if $\langle w, x\rangle \geq 0$ and $\widehat{Y}(x) = -1$ otherwise.

The algorithm goes about finding a linear separator w incrementally in a sequence of update steps.

Perceptron

- Start from the initial solution $w_0 = 0$.
- At each step $t = 0, 1, 2, \ldots$:
  - Select a random index $i \in \{1, ..., n\}$.
  - Case 1: If $y_i\langle w_t, x_i\rangle < 1$, put

$$w_{t+1} = w_t + y_i x_i\,.$$

  - Case 2: Otherwise put $w_{t+1} = w_t$.

Case 1 corresponds to what's called a *margin mistake*. The sign of the linear function may not disagree with the label, but it doesn't have the required margin that we asked for.

When an update occurs, we have

$$\langle w_{t+1}, x_i\rangle = \langle w_t, x_i\rangle + y_i\|x_i\|^2\,.$$

In this sense, the algorithm is nudging the hyperplane to be less wrong on example x_i. However, in doing so it could introduce errors on other examples. It is not yet clear that the algorithm converges to a linear separator when this is possible.

Connection to empirical risk minimization

Before we turn to the formal guarantees of the perceptron, it is instructive to see how to relate it to empirical risk minimization. In order to do so, it's helpful

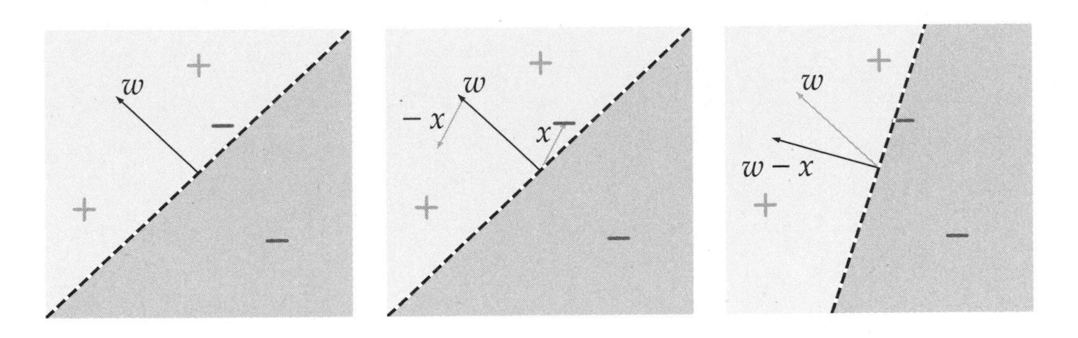

Figure 3.2: Illustration of the perceptron update. Left: One misclassified example x. Right: After update.

to introduce two *hyperparameters* to the algorithm by considering the alternative update rule:

$$w_{t+1} = \gamma w_t + \eta y_i x_i \,.$$

Here η is a positive scalar called a *learning rate* and $\gamma \in [0,1]$ is called the *forgetting rate*.

First, it's clear from the description that we're looking for a linear separator. Hence, our function class is the set of linear functions $f_w(x) = \langle w, x \rangle$, where $w \in \mathbb{R}^d$. We will sometimes call the vector w the *weight vector* or vector of *model parameters*.

An optimization method that picks a random example at each step and makes a local improvement to the model parameters is the *stochastic gradient method*. This method will figure prominently in our chapter on optimization as it is the workhorse of many machine learning applications today. The local improvement the method picks at each step is given by a local linear approximation of the loss function around the current model parameters. This linear approximation can be written neatly in terms of the vector of first derivatives, called *gradient*, of the loss function with respect to the current model parameters.

The formal update rule reads

$$w_{t+1} = w_t - \eta \nabla_{w_t} loss(f_{w_t}(x_i), y_i) \,.$$

Here, the example (x_i, y_i) is randomly chosen and the expression $\nabla_{w_t} loss(f_{w_t}(x_i), y_i)$ is the gradient of the loss function with respect to the model parameters w_t on the example (x_i, y_i). We will typically drop the vector w_t from the subscript of the gradient when it's clear from the context. The scalar $\eta > 0$ is a step size parameter that we will discuss more carefully later. For now, think of it as a small constant.

It turns out that we can connect this update rule with the perceptron algorithm by choosing a suitable loss function. Consider the loss function

$$loss(\langle w, x \rangle, y) = \max\left\{1 - y\langle w, x \rangle,\, 0\right\}.$$

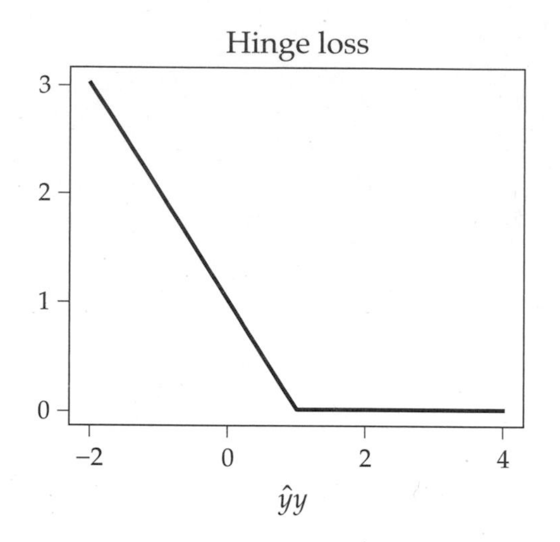

Figure 3.3: Hinge loss

This loss function is called *hinge loss*. Note that its gradient is $-yx$ when $y\langle w, x\rangle < 1$ and 0 when $y\langle w, x\rangle > 1$.

The gradient of the hinge loss is not defined when $y\langle w, x\rangle = 1$. In other words, the loss function is not differentiable everywhere. This is why technically speaking the stochastic gradient method operates with what is called a *subgradient*. The mathematical theory of subgradient optimization rigorously justifies calling the gradient 0 when $y\langle w, x\rangle = 1$. We will ignore this technicality throughout the book.

We can see that the hinge loss gives us part of the update rule in the perceptron algorithm. The other part comes from adding a weight penalty $\frac{\lambda}{2}\|w\|^2$ to the loss function that discourages the weights from growing out of bounds. This weight penalty is called *ℓ_2-regularization*, *weight decay*, or *Tikhonov regularization* depending on which field you work in. The purpose of regularization is to promote generalization. We will therefore return to regularization in detail when we discuss generalization in more depth. For now, note that the margin constraint we introduced is inconsequential unless we penalize large vectors. Without the weight penalty we could simply scale up any linear separator until it separates the points with the desired margin.

Putting the two loss functions together, we get the ℓ_2-regularized empirical risk minimization problem for the hinge loss:

$$\frac{1}{n}\sum_{i=1}^{n}\max\left\{1 - y_i\langle w, x_i\rangle, 0\right\} + \frac{\lambda}{2}\|w\|_2^2 .$$

The perceptron algorithm corresponds to solving this empirical risk objective with the stochastic gradient method. The constant η, which we dubbed the learning rate, is the step size of the stochastic gradient methods. The forgetting rate constant γ is equal to $(1-\eta\lambda)$. The optimization problem is also known as *support vector machine* and we will return to it later on.

A word about surrogate losses

When the goal was to maximize the accuracy of a predictor, we mathematically solved the risk minimization problem with respect to the *zero-one loss*

$$loss(\widehat{y}, y) = \mathbb{1}\{\widehat{y} \neq y\}$$

that gives us penalty 1 if our label is incorrect, and penalty 0 if our predicted label $\widehat{y}$ matches the true label y. We saw that the optimal predictor in this case was a *maximum a posteriori* rule, where we selected the label with higher posterior probability.

Why don't we directly solve empirical risk minimization with respect to the zero-one loss? The reason is that the empirical risk with the zero-one loss is computationally difficult to optimize directly. In fact, this optimization problem is NP-hard even for linear prediction rules.[37] To get a better sense of the difficulty, convince yourself that the stochastic gradient method, for example, fails entirely on the zero-one loss objective. Of course, the stochastic gradient method is not the only learning algorithm.

The hinge loss therefore serves as a *surrogate loss* for the zero-one loss. We hope that by optimizing the hinge loss, we end up optimizing the zero-one loss as well. The hinge loss is not the only reasonable choice. There are numerous loss functions that approximate the zero-one loss in different ways.

- The *hinge loss* is $\max\{1 - y\widehat{y}, 0\}$ and *support vector machine* refers to empirical risk minimization with the hinge loss and ℓ_2-regularization. This is what the perceptron is optimizing.
- The *squared loss* is given by $\frac{1}{2}(y - \widehat{y})^2$. Linear least-squares regression corresponds to empirical risk minimization with the squared loss.
- The *logistic loss* is $-\log(\sigma(\widehat{y}))$ when $y = 1$ and $-\log(1 - \sigma(\widehat{y}))$ when $y = -1$, where $\sigma(z) = 1/(1 + \exp(-z))$ is the logistic function. *Logistic regression* corresponds to empirical risk minimization with the logistic loss and linear functions.

Sometimes we can theoretically relate empirical risk minimization under a surrogate loss to the zero-one loss. In general, however, these loss functions are used heuristically and practitioners evaluate performance by trial and error.

Formal guarantees for the perceptron

We saw that the perceptron corresponds to finding a linear predictor using the stochastic gradient method. What we haven't seen yet is a proof that the

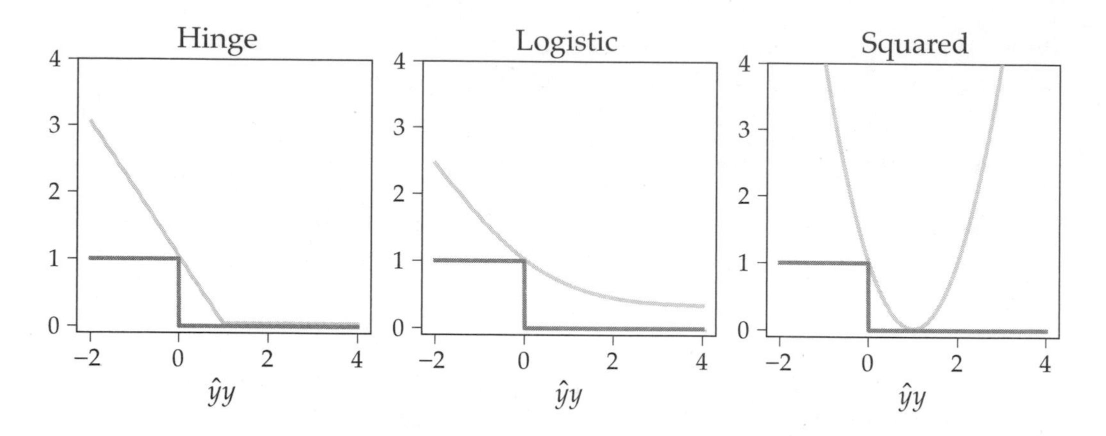

Figure 3.4: Hinge, squared, logistic loss compared with the zero-one loss

perceptron method works, and under what conditions. Recall that there are two questions we need to address. The first is why the perceptron method successfully fits the training data, a question about *optimization*. The second is why the solution should also correctly classify unseen examples drawn from the same distribution, a question about *generalization*. We will address each in turn with results from the 1960s. Even though the analysis here is over 50 years old, it has all of the essential parts of more recent theoretical arguments in machine learning.

Mistake bound

To see why we perform well on the training data, we use a *mistake bound* due to Novikoff.[38] The bound shows that if there exists a linear separator of the training data, then the perceptron will find it quickly provided that the *margin* of the separating hyperplane isn't too small.

Margin is a simple way to evaluate how well a predictor separates data. Any vector $w \in \mathbb{R}^d$ defines a hyperplane $\mathcal{H}_w = \{x : w^T x = 0\}$. Suppose that the hyperplane $\mathcal{H}_w$ corresponding to the vector w perfectly separates the data in S. Then we define the margin of such a vector w as the smallest distance of our data points to this hyperplane:

$$\gamma(S, w) = \min_{1 \le i \le n} \operatorname{dist}(x_i, \mathcal{H}_w)\,.$$

Here,

$$\operatorname{dist}(x, \mathcal{H}_w) = \min\{\|x - x'\| : x' \in \mathcal{H}_w\} = \frac{|\langle x, w\rangle|}{\|w\|}.$$

Overloading terminology, we define the margin of a *dataset* to be the maximum margin achievable by any predictor w:

$$\gamma(S) = \max_{\|w\|=1} \gamma(S, w)\,.$$

We will now show that when a dataset has a large margin, the perceptron algorithm will find a separating hyperplane quickly.

Let's consider the simplest form of the perceptron algorithm. We initialize the algorithm with $w_0 = 0$. The algorithm proceeds by selecting a random index i_t at step t, checking whether $y_{i_t} w_t^T x_{i_t} < 1$. We call this condition a margin mistake, i.e., the prediction $w_t^T x_{i_t}$ is either wrong or too close to the hyperplane. If a margin mistake occurs, the perceptron performs the update

$$w_{t+1} = w_t + y_{i_t} x_{i_t} \,.$$

That is, we rejigger the hyperplane to be more aligned with the signed direction of the mistake. If no margin mistake occurs, then $w_{t+1} = w_t$.

To analyze the perceptron we need one additional definition. Define the diameter of a data S to be

$$D(S) = \max_{(x,y) \in S} \|x\| \,.$$

We can now summarize a worst-case analysis of the perceptron algorithm with the following theorem.

Theorem 1. *The perceptron algorithm makes at most* $(2 + D(S)^2)\gamma(S)^{-2}$ *margin mistakes on any sequence of examples S that can be perfectly classified by a linear separator.*

Proof. The main idea behind the proof of this theorem is that since w only changes when you make a mistake, we can upper bound and lower bound w at each time a mistake is made, and then, by comparing these two bounds, compute an inequality on the total number of mistakes.

To find an upper bound, suppose that at step t the algorithm makes a margin mistake. We then have the inequality:

$$\begin{aligned} \|w_{t+1}\|^2 &= \|w_t + y_{i_t} x_{i_t}\|^2 \\ &= \|w_t\|^2 + 2y_{i_t}\langle w_t, x_{i_t}\rangle + \|x_{i_t}\|^2 \\ &\le \|w_t\|^2 + 2 + D(S)^2 \,. \end{aligned}$$

The final inequality uses the fact that $y_{i_t}\langle w_t, x_{i_t}\rangle < 1$. Now, let m_t denote the total number of mistakes made by the perceptron in the first t iterations. Summing up the above inequality over all the mistakes we make and using the fact that $\|w_0\| = 0$, we get our upper bound on the norm of w_t:

$$\|w_t\| \le \sqrt{m_t(2 + D(S)^2)} \,.$$

Working toward a lower bound on the norm of w_t, we will use the following argument. Let w be any unit vector that correctly classifies all points in S. If we make a mistake at iteration t, we have

$$\langle w, w_{t+1} - w_t \rangle = \langle w, y_{i_t} x_{i_t} \rangle = \frac{|\langle w, x_{i_t} \rangle|}{\|w\|} \geq \gamma(S, w) .$$

Note that the second equality here holds because w correctly classifies the point (x_{i_t}, y_{i_t}). This is where we use the assumption that the data are linearly separable. The inequality follows from the definition of margin.

Now, let $w_\star$ denote the hyperplane that achieves the maximum margin $\gamma(S)$. Instantiating the previous argument with $w_\star$, we find that

$$\|w_t\| \geq \langle w_\star, w_t \rangle = \sum_{k=1}^{t} w_\star^T (w_k - w_{k-1}) \geq m_t \gamma(S) ,$$

where the equality follows from a telescoping sum argument.

This yields the desired lower bound on the norm of w_t. Combined with the upper bound we already derived, it follows that the total number of mistakes has the bound

$$m_t \leq \frac{2 + D(S)^2}{\gamma(S)^2} . \qquad \square$$

The proof we saw has some ingredients we'll encounter again. Telescoping sums, for example, are a powerful trick used throughout the analysis of optimization algorithms. A telescoping sum lets us understand the behavior of the final iterate by decomposing it into the incremental updates of the individual iterations.

The mistake bound does not depend on the dimension of the data. This is appealing since the requirement of linear separability and high margin, intuitively speaking, become less taxing the larger the dimension is.

An interesting consequence of this theorem is that if we run the perceptron repeatedly over the same dataset, we will eventually end up with a separating hyperplane. To see this, imagine repeatedly running over the dataset until no mistake occurs on a full pass over the data. The mistake bound gives a bound on the number of passes required before the algorithm terminates.

From mistake bounds to generalization

The previous analysis shows that the perceptron finds a good predictor on the training data. What can we say about new data that we have not yet seen?

To talk about generalization, we need to make a statistical assumption about the data generating process. Specifically we assume that the data points in the training set $S = \{(x_1, y_1) \ldots, (x_n, y_n)\}$ are each drawn i.i.d. from a fixed underlying distribution $\mathcal{D}$. The labels take values $\{-1, 1\}$ and each $x_i \in \mathbb{R}^d$.

We know that the perceptron finds a good linear predictor for the training data (if it exists). What we now show is that this predictor also works on new data drawn from the same distribution.

To analyze what happens on new data, we will employ a powerful *stability* argument. Put simply, an algorithm is stable if the effect of removing or replacing a single data point is small. We will do a deep dive on stability in our chapter on generalization, but we will have a first encounter with the idea here.

The perceptron is stable because it makes a bounded number of mistakes. If we remove a data point where no mistake is made, the model doesn't change at all. In fact, it's as if we had *never seen* the data point. This lets us relate the performance on seen examples to the performance on examples in the training data on which the algorithm never updated.

Vapnik and Chervonenkis presented the following stability argument in their classic text from 1974, though the original argument is likely a decade older.[39] Their main idea was to leverage our assumption that the data are i.i.d., so we can swap the roles of training and test examples in the analysis.

Theorem 2. *Let S_n denote a training set of n i.i.d. samples from a distribution $\mathcal{D}$ that we assume has a perfect linear separator. Let $w(S)$ be the output of the perceptron on a dataset S after running until the hyperplane makes no more margin mistakes on S. Let $Z = (X, Y)$ be an additional independent sample from $\mathcal{D}$. Then, the probability of making a margin mistake on (X, Y) satisfies the upper bound*

$$\mathbb{P}[Yw(S_n)^T X < 1] \leq \frac{1}{n+1} \mathbb{E}_{S_{n+1}} \left[\frac{2 + D(S_{n+1})^2}{\gamma(S_{n+1})^2} \right] .$$

Proof. First note that

$$\mathbb{P}[Yw^T X < 1] = \mathbb{E}[\mathbb{1}\{Yw^T X < 1\}] .$$

Let $S_n = (Z_1, ..., Z_n)$ with $Z_k = (X_k, Y_k)$ and put $Z_{n+1} = Z = (X, Y)$. Note that these $n+1$ random variables are i.i.d. drawn from $\mathcal{D}$. As a purely analytical device, consider the "leave-one-out set"

$$S^{-k} = \{Z_1, \ldots, Z_{k-1}, Z_{k+1}, ..., Z_{n+1}\} .$$

Since the data are drawn i.i.d., running the algorithm on S^{-k} and evaluating it on $Z_k = (X_k, Y_k)$ is equivalent to running the algorithm on S_n and evaluating it on Z_{n+1}. These all correspond to the same random experiment and differ only in naming. In particular, we have

$$\mathbb{P}[Yw(S_n)^T X < 1] = \frac{1}{n+1} \sum_{k=1}^{n+1} \mathbb{E}[\mathbb{1}\{Y_k w(S^{-k})^T X_k < 1\}] .$$

Indeed, we're averaging quantities that are each identical to the left hand side. But recall from our previous result that the perceptron makes at most

$$m = \frac{2 + D((Z_1, \ldots, Z_{n+1}))^2}{\gamma((Z_1, \ldots, Z_{n+1}))^2}$$

margin mistakes when run on the entire sequence $(Z_1, \ldots, Z_{n+1})$. Let $i_1, \ldots, i_m$ denote the indices on which the algorithm makes a mistake in any of its cycles over the data. If $k \notin \{i_1, \ldots, i_m\}$, the output of the algorithm remains the same after we remove the kth sample from the sequence. It follows that such k satisfy $Y_k w(S^{-k}) X_k \geq 1$ and therefore k does not contribute to the summation above. The other terms can at most contribute 1 to the summation. Hence,

$$\sum_{k=1}^{n+1} \mathbb{1}\{Y_k w(S^{-k})^T X_k < 1\} \leq m\,,$$

and by linearity of expectation, as we hoped to show,

$$\mathbb{P}[Y w(S_n)^T X < 1] \leq \frac{\mathbb{E}[m]}{n+1}\,. \qquad \square$$

We can turn our mistake bounds into bounds on the empirical risk and risk achieved by the perceptron algorithm by choosing the loss function $loss(\langle w, x \rangle, y) = \mathbb{1}\{\langle w, x \rangle y < 1\}$. These bounds also imply bounds on the (empirical) risk with respect to the zero-one loss, since the prediction error is bounded by the number of margin mistakes.

Chapter notes

Rosenblatt developed the perceptron in 1957 and continued to publish on the topic in the years that followed.[40,41] The perceptron project was funded by the US Office of Naval Research (ONR), which jointly announced the project with Rosenblatt in a press conference in 1958, which led to the *New York Times* article we quoted from earlier. This development sparked significant interest in perceptron research throughout the 1960s.

The simple proof of the mistake bound we saw is due to Novikoff.[38] Block is credited with a more complicated contemporaneous proof.[42] Minsky and Papert attribute a simple analysis of the convergence guarantees for the perceptron to a 1961 paper by Papert.[43]

Following these developments Vapnik and Chervonenkis proved the generalization bound for the perceptron method that we saw earlier, relying on the kind of stability argument that we will return to in our chapter on generalization. The proof of Theorem 2 is available in their 1974 book.[39] Interestingly, by the 1970s, Vapnik and Chervonenkis must have abandoned the stability argument in favor of the VC-dimension.

In 1969, Minksy and Papert published their influential book *Perceptrons: An Introduction to Computational Geometry*.[44] Among other results, it showed that perceptrons fundamentally could not learn certain concepts, such as an XOR of its input bits. In modern language, linear predictors cannot learn parity functions. The results remain relevant in the statistical learning community and have been extended in numerous ways. On the other hand, pragmatic researchers realized one could just add the XOR to the feature vector and continue to use linear methods. We will discuss such feature engineering in the next chapter.

The dominant narrative in the field has it that Minsky and Papert's book curbed enthusiasm for perceptron research and their multilayer extensions, now better known as deep neural networks. In an updated edition of their book from 1988, Minsky and Papert argue that work on perceptrons had already slowed significantly by the time their book was published due to a lack of new results:

> One popular version is that the publication of our book so discouraged research on learning in network machines that a promising line of research was interrupted. Our version is that progress had already come to a virtual halt because of the lack of adequate basic theories.

On the other hand, the pattern recognition community had realized that perceptrons were just one way to implement linear predictors. Highleyman was arguably the first to propose empirical risk minimization and applied this technique to optical character recognition.[31] Active research in the 1960s showed how to find linear rules using linear programming techniques.[45] Aizerman, Braverman, and Rozonoer developed iterative methods to fit nonlinear rules to data.[46] All of this work was covered in depth in the first edition of Duda and Hart, which appeared five years after *Perceptrons*.

It was at this point that the artificial intelligence community first split from the pattern recognition community. While the artificial intelligence community turned toward more *symbolic* techniques in 1970s, work on statistical learning continued in Soviet and IEEE journals. The modern view of empirical risk minimization, with which we began this chapter, came out of this work and was codified by Vapnik and Chervonenkis in the 1970s.

It wasn't until the 1980s that work on pattern recognition, and with it the tools of the 1960s and earlier, took a stronger foothold in the machine learning community again.[11] We will continue this discussion in our chapter on datasets and machine learning benchmarks, which were pivotal in the return of pattern recognition to the forefront of machine learning research.

Chapter 4

Representations and Features

The starting point for prediction is the existence of a vector x from which we can predict the value of y. In machine learning, each component of this vector is called a *feature*. We would like to find a set of features that are good for prediction. Where do features come from in the first place?

In much of machine learning, the feature vector x is considered to be given. However, features are not handed down from first principles. They have to be constructed somehow, often based on models that incorporate assumptions, design choices, and human judgments. The construction of features often follows human intuition and domain-specific expertise. Nonetheless, there are several principles and recurring practices we will highlight in this chapter.

Feature representations must balance many demands. First, at a population level, they must admit decision boundaries with low error rates. Second, we must be able to optimize the empirical risk efficiently given the current set of features. Third, the choice of features influences the generalizability of the resulting model.

There are a few core patterns in feature engineering that are used to meet this set of requirements. First, there is the process of turning a measurement into a vector on a computer, which is accomplished by *quantization and embedding*. Second, in an attempt to focus on the most discriminative directions, the binned vectors are sorted relative to their similarity to a set of likely patterns through a process of *template matching*. Third, as a way to introduce robustness to noise or reduce and standardize the dimension of data, feature vectors are compressed into a low, fixed dimension via *histograms and counts*. Finally, *nonlinear liftings* are used to enable predictors to approximate complex, nonlinear decision boundaries. These processes are often iterated, and oftentimes the feature generation process itself is tuned on the collected data.

Measurement

Before we go into specific techniques and tricks of the trade, it's important to recognize the problem we're dealing with in full generality. Broadly speaking, the first step in any machine learning process is to numerically represent objects in the real world and their relationships in a way that can be processed by computers.

There is an entire scientific discipline, called measurement theory, devoted to this subject. The field of measurement theory distinguishes between a measurement procedure and the target *construct* that we wish to measure.[47,48,49] Physical temperature, for example, is a construct and a thermometer is a measurement device. Mathematical ability is another example of a construct; a math exam can be thought of as a procedure for measuring this construct. While we take reliable measurement of certain physical quantities for granted today, other measurement problems remain difficult.

It is helpful to frame feature creation as measurement. All data stems from some measurement process that embeds and operationalizes numerous important choices.[50] Measurement theory has developed a range of techniques over the decades. In fact, many measurement procedures themselves involve statistical models and approximations that blur the line between what is a feature and what is a model. Practitioners of machine learning should consult with experts on measurement within specific domains before creating ad hoc measurement procedures. More often than not there is much relevant scholarship on how to measure various constructs of interest. When in doubt it's best to choose constructs with an established theory.

Human subjects

The complexities of measurement are particularly apparent when our features are about human subjects. Machine learning problems relating to human subjects inevitably involve features that aim to quantify a person's traits, inclinations, abilities, and qualities. Researchers often try to get at these constructs by designing surveys, tests, or questionnaires. However, much data about humans is collected in an ad hoc manner, lacking clear measurement principles. This is especially true in a machine learning context.

Featurization of human subjects can have serious consequences. Recall the example of using prediction in the context of the criminal justice system. The COMPAS recidivism risk score is trained on survey questions designed using psychometric models to capture archetypes of people that might indicate future criminal behavior. The exam asks people to express their degree of agreement with statements such as "I always practice what I preach," "I have played sick to get out of something," and "I've been seen by others as cold and unfeeling."[22] Though COMPAS features have been used to predict recidivism, they have been shown to be no more predictive than using age, gender, and past criminal activity.[51]

Machine learning and data creation involving human subjects should be ethically evaluated in the same manner as any other scientific investigation with humans. Depending on context, different ethical guidelines and regulations exist that aim at protecting human research subjects. The 1979 Belmont Report is one ethical framework, commonly applied in the United States. It rests on the three principles of respect for persons, beneficence, and justice. Individuals should be treated as autonomous agents. Harm should be minimized, while benefits should be maximized. Inclusion and exclusion should not unduly burden specific individuals, as well as marginalized and vulnerable groups.

Universities typically require obtaining institutional approval and detailed training before conducting human subject research. These rules apply also when data is collected from and about humans online.

We advise readers to familiarize themselves with all applicable rules and regulations regarding human subject research at their institution.

Quantization

Signals in the real world are often continuous and we have to choose how to discretize them for use in a machine learning pipeline. Broadly speaking, such practices fall under the rubric of *quantization*. In many cases, our goal is to quantize signals so that we can reconstruct them almost perfectly. This is the case with high-resolution photography, high-fidelity analog-to-digital conversion of audio, and perfect sequencing of proteins. In other cases, we may want to record only skeletons of signals that are useful for particular tasks. This is the case with almost all quantizations of human beings. While we do not aim to do a thorough coverage of this subject, we note quantization is an essential preprocessing step in any machine learning pipeline. Improved quantization schemes may very well translate into improved machine learning applications. Let us briefly explore a few canonical examples and issues of quantization in contemporary data science.

Images

Consider the raw bitmap formatting of an image. A bitmap file is an array indexed by three coordinates. Mathematically, this corresponds to a *tensor* of order 3. The first two coordinates index space and the third indexes a color channel. That is, x_{ijk} denotes the intensity at row i, column j, and color channel k. This representation summarizes an image by dividing two-dimensional space into a regular grid, and then counting the quantity of each of three primary colors at each grid location.

This pixel representation suffices to render an image on a computer screen. However, it might not be useful for prediction. Images of the same scene with different lighting or photographic processes might end up being quite dissimilar in a pixel representation. Even small translations of two images might be far apart from each other in a pixel representation. Moreover, from the vantage point of linear classification, we could train a linear predictor on any ordering of the pixels, but scrambling the pixels in an image certainly makes it unrecognizable. We will describe transformations that address such shortcomings of the pixel representation in the sequel.

Text

Consider a corpus of n documents. These documents will typically have varying length and vocabulary. To embed a document as a vector, we can create a giant vector for every word in the document where each component of the vector corresponds to one dictionary word. The dimension of the vector is therefore the size of the dictionary. For each word that occurs in the document we set the corresponding coordinate of the vector to 1. All other coordinates corresponding to words not in the document we set to 0.

This is called a *one-hot encoding* of the words in the dictionary. The one-hot encoding might seem both wasteful due to the high dimension and lossy since we don't encode the order of the words in the document. Nonetheless, it turns out to be useful in a number of applications. Since typically the language has more dictionary words than the length of the document, the encoding maps a document to a very sparse vector. A corpus of documents would map to a set of sparse vectors.

Template matching

While quantization can often suffice for prediction problems, we highlighted above how this might be too fine a representation to encode when data points are similar or dissimilar. Oftentimes there are higher-level patterns that might be more representative for discriminative tasks. A popular way to extract these patterns is *template matching*, where we extract the correlation of a feature vector x with a known pattern v, called *template*.

At a high level, a template match creates a feature x' from the feature x by binning the correlation with a template. A simple example would be to fix a template v and compute

$$x' = \max\left\{v^T x, 0\right\}.$$

We now describe some more specific examples that are ubiquitous in pattern classification.

Fourier, cosine, and wavelet transforms

One of the foundational patterns that we match to spatial or temporal data is the sinusoid. Consider a vector in $\mathbb{R}^d$ and the transformation with kth component

$$x'_k = |v_k^T x|\,.$$

Here the ℓth component of v_k is given by $v_{k\ell} = \exp(2\pi i k\ell/d)$. In this case we are computing the magnitude of the *Fourier transform* of the feature vector. This transformation measures the amount of oscillation in a vector. The magnitude of the Fourier transform has the following powerful property. Suppose z is a translated version of x so that

$$z_k = x_{(k+s) \bmod d}$$

for some shift s. Then one can check that for any v_k,

$$|v_k^T x| = |v_k^T z|\,.$$

That is, the magnitude of the Fourier transform is *translation invariant*. There are a variety of other transforms that fall into this category of capturing the transformation invariant content of signals, including cosine and wavelet transforms.

Convolution

For spatial or temporal data, we often consider two data points to be similar if we can translate one to align with another. For example, small translations of digits are certainly the same digit. Similarly, two sound utterances delayed by a few milliseconds are the same for most applications. *Convolutions* are small templates that are translated over a feature figure to count the number of occurrences of some pattern. The output of a convolution will have the same spatial extent as the input, but will be a "heat map" denoting the amount of correlation with the template at each location in the vector. Multiple convolutions can be concatenated to add discriminative power. For example, if one wants to design a system to detect animals in images, one might design a template for heads, legs, and tails, and then linear combinations of these appearances might indicate the existence of an animal.

Summarization and histograms

Histograms summarize statistics about counts in data. These serve as a method for both reducing the dimensionality of an input and removing noise in the data. For instance, if a feature vector was the temperature in a location over a year, this could be converted into a histogram of temperatures that might better discriminate between locations. As another example, we could downsample an image by making a histogram of the amount of certain colors in local regions of the image.

Bag of words

We could summarize a piece of text by summing up the one-hot encoding of each word that occurs in the text. The resulting vector would have entries where each component is the number of times the associated word appears in the document. This is a *bag of words* representation of the document. A related representation that might take the structure of text better into account might have a bag of words for every paragraph or some shorter-length contiguous context.

Bag of words representations are surprisingly powerful. For example, documents about sports tend to have a different vocabulary than documents about fashion, and hence are far away from each other in such an embedding. Since the number of unique words in any given document is much less than all possible words, bag-of-words representations can be reasonably compact and sparse. The representations of text as large-dimensional sparse vectors can be deployed for predicting topics and sentiment.

Downsampling/pooling

Another way to summarize and reduce dimension is to *locally* average a signal or image. This is called downsampling. For example, we could break an image into 2 x 2 grids, and take the average or maximum value in each grid. This would reduce the image size by a factor of 4, and would summarize local variability in the image.

Nonlinear predictors

Once we have a feature vector x that we feel adequately compresses and summarizes our data, the next step is building a prediction function $f(x)$. The simplest such predictors are linear functions of x, and linear functions are quite powerful: all of the transformations we have thus far discussed in this chapter often suffice to arrange data such that linear decision rules have high accuracy.

However, we oftentimes desire further expressivity brought by more complex decision rules. We now discuss many techniques that can be used to build such nonlinear rules. Our emphasis will highlight how most of these operations can be understood as embedding data in spaces where linear separation is possible. That is: we seek a nonlinear transformation of the feature vector so that linear prediction works well on the transformed features.

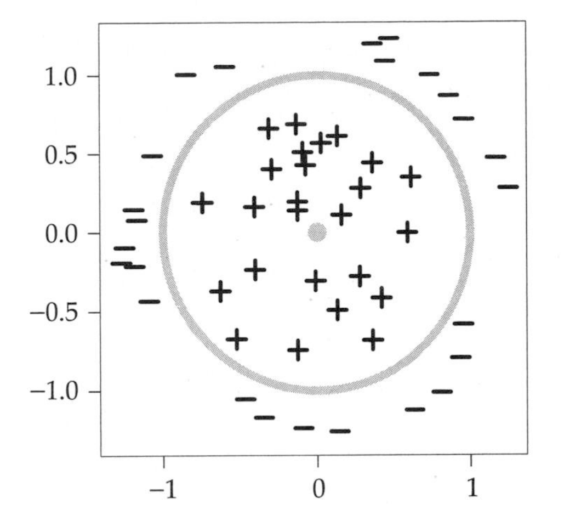

Figure 4.1: An easy problem for polynomial classification. Here, the large gray dot denotes the center of the displayed circle.

Polynomials

Polynomials are simple and natural nonlinear predictors. Consider the dataset in the figure below. Here the data clearly can't be separated by a linear function, but a quadratic function would suffice. Indeed, we'd just use the rule that if $(x_1 - c_1)^2 + (x_2 - c_2)^2 \leq c_3$ then (x_1, x_2) would have label $y = 1$. This rule is a quadratic function of (x_1, x_2).

To fit quadratic functions, we only need to fit the *coefficients* of the function. Every quadratic function can be written as a sum of quadratic monomials. This means that we can write quadratic function estimation as fitting a *linear* function to the feature vector:

$$\Phi_2^{\text{poly}}(x) = \begin{bmatrix} 1 & x_1 & x_2 & x_1^2 & x_1 x_2 & x_2^2 \end{bmatrix}^T$$

Any quadratic function can be written as $w^T \Phi_2^{\text{poly}}(x)$ for some w. The map Φ_2^{poly} is a *lifting function* that transforms a set of features into a more expressive set of features.

The features associated with the quadratic polynomial lifting function have an intuitive interpretation as cross products of existing features. The resulting prediction function is a linear combination of pairwise products of features. Hence, these features capture co-occurrence and correlation of a set of features.

The number of coefficients of a generic quadratic function in d dimensions is

$$\binom{d+2}{2},$$

which grows quadratically with dimension. For general degree p polynomials, we could construct a lifting function $\Phi_p^{\text{poly}}(x)$ by listing all of the monomials with degree at most p. Again, any polynomial of degree p can be written as $w^T \Phi_p^{\text{poly}}(x)$ for some w. In this case, $\Phi_p^{\text{poly}}(x)$ would have

$$\binom{d+p}{p}$$

terms, growing roughly as d^p. It shouldn't be too surprising to see that as the degree of the polynomial grows, increasingly complex behavior can be approximated.

How many features do you need?

Our discussion of polynomials began with the motivation of creating nonlinear decision boundaries. But we saw that we could also view polynomial boundaries as taking an existing feature set and performing a nonlinear transformation to embed that set in a higher dimensional space where we could then search for a linear decision boundary. This is why we refer to nonlinear feature maps as *lifts*.

Given expressive enough functions, we can always find a lift such that a particular dataset can be mapped to any desired set of labels. How high of a dimension is necessary? To gain insights into this question, let us stack all of the data points $x_1, \ldots, x_n \in \mathbb{R}^d$ into a matrix X with n rows and d columns. The predictions across the entire dataset can now be written as

$$\hat{y} = Xw\,.$$

If the x_i are linearly independent, then as long as $d \geq n$, we can make *any* vector of predictions y by finding a corresponding vector w. For the sake of *expressivity*, the goal in feature design will be to find lifts into high dimensional space such that our data matrix X has linearly independent columns. This is one reason why machine learning practitioners lean toward models with more parameters than data points. Models with more parameters than data points are called *overparameterized*.

As we saw in the analysis of the perceptron, the key quantities that governed the number of mistakes in the perceptron algorithm were the maximum norm of x_k and the norm of the optimal w. Importantly, the dimension of the data played no role. Designing features where w has controlled norm is a domain-specific challenge, but has nothing to do with dimension. As we will see in the remainder of this book, high-dimensional models have many advantages and few disadvantages in the context of prediction problems.

Basis functions

Polynomials are an example of *basis functions*. More generally, we can write prediction functions as linear combinations of B general nonlinear functions $\{b_k\}$:

$$f(x) = \sum_{k=1}^{B} w_k b_k(x) \,.$$

In this case, there is again a lifting function $\Phi_{\text{basis}}(x)$ into B dimensions where the kth component is equal to $b_k(x)$ and $f(x) = w^T \Phi_{\text{basis}}(x)$. There are a variety of basis functions used in numerical analysis including trigonometric polynomials, spherical harmonics, and splines. The basis function most suitable for a given task is often dictated by prior knowledge in the particular application domain.

Radial basis functions are particularly useful in pattern classification. A radial basis function has the form

$$b_z(x) = \phi(\|x - z\|)$$

where $z \in \mathbb{R}^d$ and $\phi : \mathbb{R} \to \mathbb{R}$. Most commonly,

$$\phi(t) = e^{-\gamma t^2}$$

for some $\gamma > 0$. In this case, given $z_1, \ldots, z_k$, our functions take the form

$$f_k(x) = \sum_{j=1}^{k} w_j e^{-\gamma \|x - z_j\|^2} \,.$$

Around each anchor point z_k, we place a small Gaussian bump. Combining these bumps with different weights allows us to construct arbitrary functions.

How to choose the z_k? In low-dimensions, z_k could be placed on a regular grid. But the number of bases would then need to scale exponentially with dimension. In higher dimensions, there are several other possibilities. One is to use the set of training data. This is well motivated by the theory of *reproducing kernels*. Another option would be to pick the z_k at random, inducing *random features*. A third idea would be to search for the best z_i. This motivates our study of *neural networks*. As we will now see, all three of these methods are powerful enough to approximate any desired function, and they are intimately related to each other.

Kernels

One way around high dimensionality is to constrain the space of prediction function to lie in a low-dimensional subspace. Which subspace would be

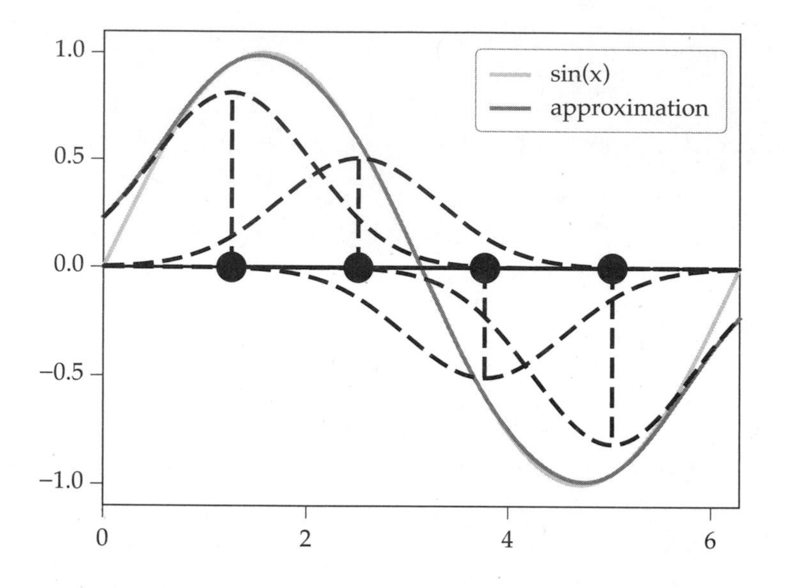

Figure 4.2: Radial basis function approximation of $\sin(x)$ with four Gaussian bumps that sum to approximate the function

useful? In the case of linear functions, there is a natural choice: the span of the training data. By the fundamental theorem of linear algebra, any vector in $\mathbb{R}^d$ can be written as a sum of a vector in the span of the training data and a vector orthogonal to all of the training data. That is,

$$w = \sum_{i=1}^{n} \alpha_i x_i + v$$

where v is orthogonal to the x_i. But if v is orthogonal to every training data point, then in terms of prediction,

$$w^T x_i = \sum_{j=1}^{n} \alpha_j x_j^T x_i \,.$$

That is, the v has no bearing on in-sample prediction whatsoever. Also note that prediction is only a function of the dot products between the training data points. In this section, we consider a family of prediction functions built with such observations in mind: we will look at functions that expressly let us compute dot products between liftings of points, noting that our predictions will be linear combinations of such lifted dot products.

Let $\Phi(x)$ denote any lifting function. Then

$$k(x, z) := \Phi(x)^T \Phi(z)$$

is called the *kernel function* associated with the feature map Φ. Such kernel functions have the property that for any $x_1, \ldots, x_n$, the matrix K with entries

$$K_{ij} = k(x_i, x_j)$$

is positive semidefinite. This turns out to be the key property to define kernel functions. We say a symmetric function $k : \mathbb{R}^d \times \mathbb{R}^d \to \mathbb{R}$ is a kernel function if it has this positive semidefiniteness property.

It is clear that positive combinations of kernel functions are kernel functions, since this is also true for positive semidefinite matrices. It is also true that if k_1 and k_2 are kernel functions, then so is $k_1 k_2$. This follows because the element-wise product of two positive semidefinite matrices is positive semidefinite.

Using these rules, we see that the function

$$k(x,z) = (a + bx^T z)^p \,,$$

where $a, b \geq 0$, p a positive integer, is a kernel function. Such kernels are called polynomial kernels. For every polynomial kernel, there exists an associated lifting function Φ with each coordinate of Φ being a monomial such that

$$k(x,z) = \Phi(x)^T \Phi(z) \,.$$

As a simple example, consider the one-dimensional case of a kernel

$$k(u,v) = (1 + uv)^2 \,.$$

Then $k(u,v) = \Phi(u)^T \Phi(v)$ for

$$\Phi(u) = \begin{bmatrix} 1 \\ \sqrt{2}u \\ u^2 \end{bmatrix} \,.$$

We can generalize polynomial kernels to *Taylor series* kernels. Suppose that the one dimensional function h has a convergent Taylor series for all $t \in [-R, R]$:

$$h(t) = \sum_{j=1}^{\infty} a_j t^j$$

where $a_j \geq 0$ for all j. Then the function

$$k(x,z) = h(\langle x, z \rangle)$$

is a positive definite kernel. This follows because each term $\langle x, z \rangle^j$ is a kernel, and we are taking a nonnegative combination of these polynomial kernels. The feature space of this kernel is the span of the monomials of degrees where the a_j are nonzero.

Two example kernels of this form are the *exponential kernel*

$$k(x,z) = \exp(\gamma \langle x, z \rangle)$$

and the *arcsine kernel*

$$k(x,z) = \sin^{-1}(\langle x, z \rangle) \,,$$

which is a kernel for x, z on the unit sphere.

Another important kernel is the Gaussian kernel:

$$k(x,z) = \exp\left(-\tfrac{\gamma}{2}\|x-z\|^2\right).$$

The Gaussian kernel can be thought of as first lifting data using the exponential kernel then projecting onto the unit sphere in the lifted space.

We note that there are many kernels with the same feature space. Any Taylor Series kernel with positive coefficients will have the same set of features. The feature space associated Gaussian kernel is equivalent to the span of radial basis functions. What distinguishes the kernels beyond the features they represent? The key is to look at the norm. Suppose we want to find a fit of the form

$$f(x_j) = w^T\Phi(x_j) \qquad \text{for } j = 1,2,\ldots,n.$$

In the case when Φ maps into a space with more dimensions than the number of data points we have acquired, there will be an infinite number of w vectors that perfectly interpolate the data. As we saw in our introduction to supervised learning, a convenient means to pick an interpolator is to choose the one with the smallest norm. Let's see how the norm interacts with the form of the kernel. Suppose our kernel is a Taylor series kernel

$$h(t) = \sum_{j=1}^{\infty} a_j\langle x,z\rangle^j.$$

Then the smaller the a_j, the larger the corresponding w_j should have to be. Thus, the a_j in the kernel expansion govern how readily we allow each feature in a least-norm fit. If we consider the exponential kernel with parameter γ, then $a_j = \frac{1}{j!}\gamma^j$. Hence, for large values of γ, only low degree terms will be selected. As we decrease γ, we allow for higher degree terms to enter the approximation. Higher degree terms tend to be more sensitive to perturbations in the data than lower degree ones, so γ should be set as large as possible while still providing desirable predictive performance.

The main appeal of working with kernel representations is they translate into simple algorithms with bounded complexity. Since we restrict our attention to functions in the span of the data, our functions take the form

$$f(x) = \left(\sum_i \alpha_i\Phi(x_i)^T\right)\Phi(x) = \sum_i \alpha_i k(x_i,x).$$

We can thus pose all of our optimization problems in terms of the coefficients α_i. This means that any particular problem will have at most n parameters to search for. Even the norm of f can be computed without ever explicitly computing the feature embedding. Recall that when $f(x) = w^T\Phi(x)$ with $w = \sum_i \alpha_i\Phi(x_i)$, we have

$$\|w\|^2 = \left\|\sum_i \alpha_i\Phi(x_i)\right\|^2 = \alpha^T K\alpha,$$

where K is the matrix with ijth entry $k(x_i, x_j)$. As we will see in the optimization chapter, such representations turn out to be optimal in most machine learning problems. Functions learned by ERM methods on kernel spaces are weighted sums of the similarity (dot product) between the training data and the new data. When k is a Gaussian kernel, this relationship is even more evident: the optimal solution is simply a radial basis function whose anchor points are given by the training data points.

Neural networks

Though the name originates from the study of neuroscience, modern neural nets arguably have little to do with the brain. Neural nets are mathematically a composition of differentiable functions, typically alternating between componentwise nonlinearities and linear maps. The simplest example of a neural network would be

$$f(x) = w^T \sigma(Ax + b)$$

where w and b are vectors, A is a matrix, and σ is a componentwise nonlinearity applying the same nonlinear function to each component of its input.

The typically used nonlinearities are not Gaussian bumps. Indeed, until recently most neural nets used *sigmoid nonlinearities* where $\sigma(t)$ is some function that is 0 at negative infinity, 1 at positive infinity, and strictly increasing. Popular choices of such functions include $\sigma(t) = \tanh(t)$ or $\sigma(t) = \frac{1}{\exp(-t)+1}$. More recently, another nonlinearity became overwhelmingly popular, the *rectified linear unit* or ReLU:

$$\sigma(t) = \max\{t, 0\}.$$

This simple nonlinearity is easy to implement and differentiate in hardware.

Though these nonlinearities are all different, they all generate similar function spaces that can approximate each other. In fact, just as was the case with kernel feature maps, neural networks are powerful enough to approximate any continuous function if enough bases are used. A simple argument by Cybenko clarifies why only a bit of nonlinearity is needed for universal approximation.[52]

Suppose that we can approximate the unit step function $u(t) = \mathbb{1}\{t > 0\}$ as a linear combination of shifts of $\sigma(t)$. A sigmoidal function like tanh is already such an approximation and $\tanh(\alpha t)$ converges to the unit step as α approaches ∞. For ReLU functions we have for any $c > 0$ that

$$\frac{1}{2c}\max\{t+c, 0\} - \frac{1}{2c}\max\{t-c, 0\} = \begin{cases} 0 & t < -c \\ 1 & t > c \\ \frac{t+c}{2c} & \text{otherwise} \end{cases}.$$

It turns out that approximation of such step functions is all that is needed for universal approximation.

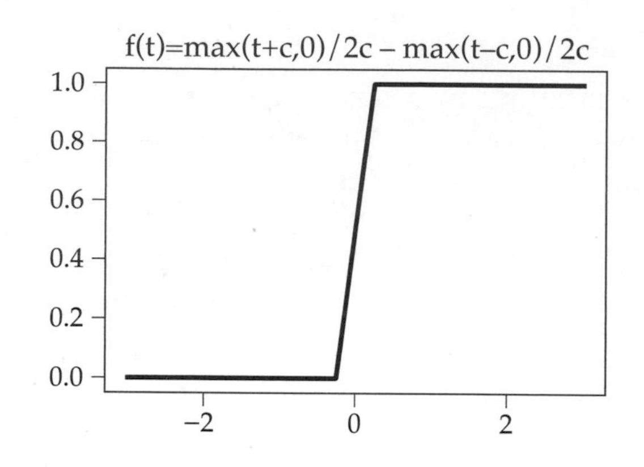

Figure 4.3: Creating a step function from ReLUs with $c = 1/4$

To see why, suppose we have a nonzero function g that is not well approximated by a sum of the form

$$\sum_{i=1}^{N} w_i \sigma(a_i^T x + b_i)\,.$$

This means that we must have a nonzero function f that lies outside the span of sigmoids. We can take f to be the projection of g onto the orthogonal complement of the span of the sigmoidal functions. This function f will satisfy

$$\int \sigma(a^T x + b) f(x)\, \mathrm{d}x = 0$$

for all vectors a and scalars b. In turn, since our nonlinearity can approximate step functions, this implies that for any a and any t_0 and t_1,

$$\int \mathbb{1}\{t_0 \le a^T x \le t_1\} f(x)\, \mathrm{d}x = 0\,.$$

We can approximate any continuous function as a sum of the indicator function of intervals, which means

$$\int h(a^T x) f(x)\, \mathrm{d}x = 0$$

for any continuous function h and any vector a. Using $h(t) = \exp(it)$ proves that the Fourier transform of f is equal to zero, and hence that f itself equals zero. This is a contradiction.

The core of Cybenko's argument is a reduction to approximation in one dimension. From this perspective, it suffices to find a nonlinearity that can well approximate "bump functions," which are nearly equal to zero outside a specified interval and are equal to 1 at the center of the interval. This opens up a variety of potential nonlinearities for use as universal approximators.

While elegant and simple, Cybenko's argument does not tell us *how many* terms we need in our approximation. More refined work on this topic was pursued in the 1990s. Barron[53] used a similar argument about step functions combined with a powerful randomized analysis due to Maurey.[54] Similar results were derived for sinusoids[55] by Jones and for ReLU networks[56] by Breiman. All of these results showed that two-layer networks sufficed for universal approximation, and quantified how the number of basis functions required scaled with respect to the complexity of the approximated function.

Random features

Though the idea seems a bit absurd, a powerful means of choosing basis functions is by random selection. Suppose we have a parametric family of basis functions $b(x;\vartheta)$. A random feature map chooses $\vartheta_1, \dots, \vartheta_D$ from some distribution on ϑ and uses the feature map

$$\Phi_{\text{rf}}(x) = \begin{bmatrix} b(x;\vartheta_1) \\ b(x;\vartheta_2) \\ \vdots \\ b(x;\vartheta_D) \end{bmatrix} .$$

The corresponding prediction functions are of the form

$$f(x) = \sum_{k=1}^{D} w_k b(x;\vartheta_k) ,$$

which looks very much like a neural network. The main difference is a matter of emphasis: here we are stipulating that the parameters ϑ_k are random variables, whereas in neural networks, ϑ_k would be considered parameters to be determined by the particular function we aim to approximate.

Why might such a random set of functions work well to approximate complex functional behavior? First, from the perspective of optimization, it might not be too surprising that a random set of nonlinear basis functions will be linearly independent. Hence, if we choose enough of them, we should be able to fit any set of desired labels.

Second, random feature maps are closely connected with kernel spaces. The connections were initially drawn out in work by Rahimi and Recht.[57,58] Any random feature map generates an empirical kernel, $\Phi_{\text{rf}}(x)^T \Phi_{\text{rf}}(z)$. The expected value of this kernel can be associated with some Reproducing Kernel Hilbert Space:

$$\begin{aligned} \mathbb{E}[\tfrac{1}{D}\Phi_{\text{rf}}(x)^T \Phi_{\text{rf}}(z)] &= \mathbb{E}\left[\frac{1}{D}\sum_{k=1}^{D} b(x;\vartheta_k) b(z;\vartheta_k)\right] \\ &= \mathbb{E}[b(x;\vartheta_1) b(z;\vartheta_1)] = \int p(\vartheta) b(x;\vartheta) b(z;\vartheta)\, \mathrm{d}\vartheta . \end{aligned}$$

In expectation, the random feature map yields a kernel given by the final integral expressions. There are many interesting kernels that can be written as such an integral. In particular, the Gaussian kernel would arise if

$$p(\vartheta) = \mathcal{N}(0, \gamma I)$$
$$b(x; \vartheta) = [\cos(\vartheta^T x),\ \sin(\vartheta^T x)]\,.$$

To see this, recall that the Fourier transform of a Gaussian is a Gaussian, and write:

$$\begin{aligned}
&k(x, z)\\
&= \exp(-\tfrac{\gamma}{2}\|x - z\|^2)\\
&= \frac{1}{(2\pi\gamma)^{d/2}} \int e^{-\frac{\|v\|^2}{2\gamma}} \exp(iv^T(x - z))\, dv\\
&= \frac{1}{(2\pi\gamma)^{d/2}} \int e^{-\frac{\|v\|^2}{2\gamma}} \left\{\cos(v^T x)\cos(v^T z) + \sin(v^T x)\sin(v^T z)\right\} dv\,.
\end{aligned}$$

This calculation gives new insights into the feature space associated with a Gaussian kernel. It shows that the Gaussian kernel is a continuous mixture of inner products of sines and cosines. The sinusoids are weighted by a Gaussian function on their frequency: high frequency sinusoids have vanishing weight in this expansion. The parameter γ controls how quickly the higher frequencies are damped. Hence, the feature space here can be thought of as low frequency sinusoids. If we sample a frequency from a Gaussian distribution, it will be low frequency (i.e., have small norm) with high probability. Hence, a random collection of low frequency sinusoids approximately spans the same space as that spanned by a Gaussian kernel.

If instead of using sinusoids, we chose our random features to be $\mathrm{ReLU}(v^T x)$, our kernel would become

$$\begin{aligned}
k(x, z) &= \frac{1}{(2\pi)^{d/2}} \int \exp\left(-\frac{\|v\|^2}{2}\right) \mathrm{ReLU}(v^T x)\,\mathrm{ReLU}(v^T z)\, dv\\
&= \|x\|\|z\| \left\{\sin(\vartheta) + (\pi - \vartheta)\cos(\vartheta)\right\}\,,
\end{aligned}$$

where

$$\vartheta = \cos^{-1}\left(\frac{\langle x, z\rangle}{\|x\|\|z\|}\right)\,.$$

This computation was first made by Cho and Saul.[59] Both the Gaussian kernel and this "ReLU kernel" are universal Taylor kernels, and, when plotted, we see even are comparable on unit norm data.

Prediction functions built with random features are just randomly wired neural networks. This connection is useful for multiple reasons. First, as we will see in the next chapter, optimization of the weights w_k is far easier than

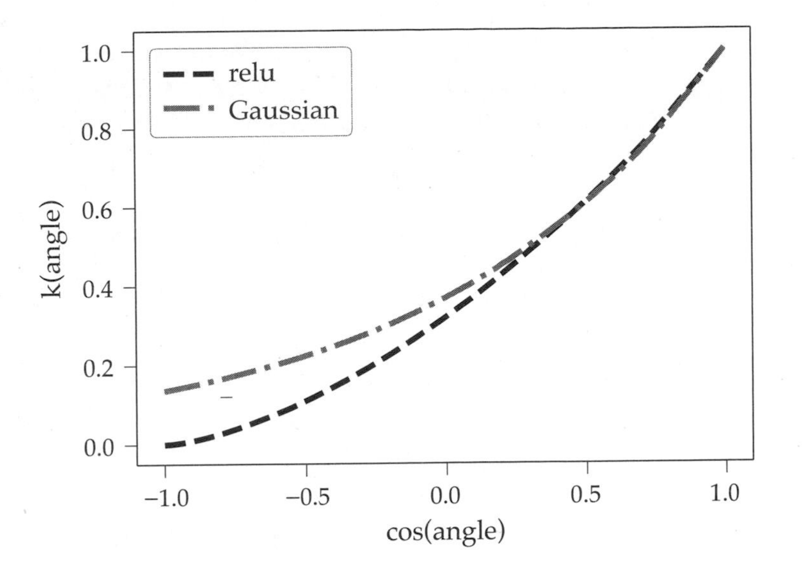

Figure 4.4: Comparison of the Gaussian and ReLU kernels as a function of the angle between two unit norm vectors

joint optimization of the weights and the parameters ϑ. Second, the connection to kernel methods makes the generalization properties of random features straightforward to analyze. Third, much of the recent theory of neural nets is based on connections between random feature maps and the randomization at initialization of neural networks. The main drawback of random features is that, in practice, they often require large dimensions before they provide predictions on par with neural networks. These trade-offs are worth considering when designing and implementing nonlinear prediction functions.

Returning to the radial basis expansion

$$f(x) = \sum_{j=1}^{k} w_j e^{-\gamma \|x - z_j\|^2},$$

we see that this expression could be a neural network, a kernel machine, or a random feature network. The main distinction between these three methods is how the z_j are selected. Neural nets require some sort of optimization procedure to find the z_j. Kernel machines place the z_j on the training data. Random features select z_j at random. The best choice for any particular prediction problem will always be dictated by the constraints of practice.

Chapter notes

To our knowledge, there is no full and holistic account of measurement and quantization, especially when quantization is motivated by data science applications. From the statistical signal processing viewpoint, we have a full and complete theory of quantization in terms of understanding what signals can be reconstructed from digital measurements. The Nyquist-Shannon theory allows us to understand what parts of signals may be lost and what artifacts are introduced. Such theory is now canon in undergraduate courses on signal processing. See, e.g., Oppenheim and Willsky.[60] For task-driven sampling, the field remains quite open. The theory of compressive sensing led to many recent and exciting developments in this space, showing that task-driven sampling where we combined domain knowledge, computation, and device design could reduce the data acquisition burdens of many pattern recognition tasks.[61] The theory of experiment design and survey design has taken some cues from task-driven sampling.

Reproducing kernels have been used in pattern recognition and machine learning for nearly as long as neural networks. Kernels and Hilbert spaces were first used in time series prediction in the late 1940s, with fundamental work by Karhunen-Loève showing that the covariance function of a time series was a Mercer Kernel.[62,63] Shortly thereafter, Reproducing Kernel Hilbert Spaces were formalized by Aronszajn in 1950.[64] Parzen was likely the first to show that time series prediction problems could be reduced to solving least squares problems in an RKHS and hence predictions could be computed by solving a linear system.[65] Wahba's survey of RKHS techniques in statistics covers many other developments post-Parzen.[66] For further reading on the theory and application of kernels in machine learning, consult the texts by Schölkopf and Smola[67] and Shawe-Taylor and Cristianini.[68]

Also since its inception, researchers have been fascinated by the approximation power of neural networks. Rosenblatt discussed properties of universal approximation in his monograph on neurodynamics.[41] It was in the 1980s when it became clear that though neural networks were able to approximate any continuous function, they needed to be made more complex and intricate in order to achieve high-quality approximation. Cybenko provided a simple proof that neural nets were dense in the space of continuous functions, though did not estimate how large such networks might need to be.[52] An elegant, randomized argument by Maurey[54] led to a variety of approximation results that quantified how many basis terms were needed. Notably, Jones showed that a simple greedy method could approximate any continuous function with a sum of sinusoids.[55] Barron shows that similar greedy methods could be used[53] to build neural nets that approximated any function. Breiman analyzed ReLU networks using the same framework.[56] The general theory of approximation by bases is rich, and Pinkus' book details some of the necessary and sufficient conditions to achieve approximations with as few bases as possible.[69]

Randomly wired neural networks also have a long history in pattern recognition. Minsky's first electronic neural network, SNARC, was randomly wired. The story of SNARC (Stochastic Neural Analog Reinforcement Calculator) is rather apocryphal. There are no known photographs of the assembled device, although a 2019 article by Akst has a picture of one of the "neurons."[70] The commonly referenced publication, a Harvard technical report, appears to not be published. However, the "randomly wired" story lives on, and it is one that Minsky told countless times through his life. Many years later, Rahimi and Recht built upon the approximation theory of Maurey, Barron, and Jones to show that random combinations of basis functions could approximate continuous functions well, and that such random combinations could be thought of as approximately solving prediction problems in an RKHS.[58,57,71] This work was later used as a means to understand neural networks, and, in particular, the influence of their random initializations. Daniely et al. computed the kernel spaces associated with that randomly initialized neural networks,[72] and Jacot et al. pioneered a line of work using kernels to understand the dynamics of neural net training.[73]

There has been noted cultural tension between the neural net and kernel "camps." For instance, the tone of the introduction of work by Decoste and Schölkopf telegraphs a disdain by neural net proponents of the Support Vector Machine.[74]

> Initially, SVMs had been considered a theoretically elegant spin-off of the general but, allegedly, largely useless VC-theory of statistical learning. In 1996, using the first methods for incorporating prior knowledge, SVMs became competitive with the state of the art in the handwritten digit classification benchmarks that were popularized in the machine learning community by AT&T and Bell Labs. At that point, practitioners who are not interested in theory, but in results, could no longer ignore SVMs.

With the rise of deep learning, however, there are a variety of machine learning benchmarks where SVMs or other kernel methods fail to match the performance of neural networks. Many have dismissed kernel methods as a framework whose time has past. However, kernels play an ever more active role in helping to better understand neural networks and insights from deep learning have helped to improve the power of kernel methods on pattern recognition tasks.[75] Neural nets and kernels are complementary, and active research in machine learning aims to bring these two views more closely together.

Chapter 5

Optimization

In Chapter 2, we devised a closed form expression for the optimal decision rule assuming we have a probability model for the data. Then we turned to empirical risk minimization (ERM), where we instead rely on numerical methods to discover good decision rules when we don't have such a probability model. In this chapter, we take a closer look at how to solve empirical risk minimization problems effectively. We focus on the core optimization methods commonly used to solve empirical risk minimization problems and on the mathematical tools used to analyze their running times.

Our main subject will be *gradient descent* algorithms and how to shape loss functions so that gradient descent succeeds. Gradient descent is an iterative procedure that iterates among possible models, at each step replacing the old model with one with lower empirical risk. We show that the class of optimization problems where gradient descent is guaranteed to find an optimal solution is the set of *convex functions*. When we turn to risk minimization, this means that gradient descent will find the model that minimizes the empirical risk whenever the loss function is convex and the decision function is a linear combination of features.

We then turn to studying *stochastic* gradient descent, the workhorse of machine learning. Stochastic gradient descent is effectively a generalization of the perceptron learning rule. Its generality enables us to apply it to a variety of function classes and loss functions and guarantee convergence even if the data may not be separable. We spend a good deal of time looking at the dynamics of the stochastic gradient method to try to motivate why it is so successful and popular in machine learning.

Starting from the convex case, we work toward more general nonconvex problems. In particular, we highlight two salient features of gradient descent and stochastic gradient descent that are particular to empirical risk minimization and help to motivate the resilience of these methods.

First, we show that even for problems that are not convex, gradient descent for empirical risk minimization has an *implicit convexity* property that encourages convergence. Though we explicitly optimize over function representations that are computationally intractable to optimize in the worst case, it turns out that we can still reason about the convergence of the predictions themselves.

Second, we show that gradient descent implicitly manages the complexity of the prediction function, encouraging solutions of low complexity in cases where infinitely many solutions exist. We close the chapter with a discussion of other methods for empirical risk minimization that more explicitly account for model complexity and stable convergence.

Optimization basics

Stepping away from empirical risk minimization for a moment, consider the general minimization problem

$$\text{minimize}_w \quad \Phi(w)$$

where $\Phi\colon \mathbb{R}^d \to \mathbb{R}$ is a real-valued function over the domain $\mathbb{R}^d$.

When and how can we minimize such a function? Before we answer this question, we need to formally define what we're shooting for.

Definition 5. *A point $w_\star$ is a* minimizer *of Φ if $\Phi(w_\star) \le \Phi(w)$ for all w. It is a* local minimizer *of Φ if for some $\epsilon > 0$, $\Phi(w_\star) \le \Phi(w)$ for all w such that $\|w - w_\star\| \le \epsilon$. Sometimes we will refer to minimizers as* global minimizers *to contrast against local minimizers.*

The figure below presents example functions and their minima. In the first illustration, there is a unique minimizer. In the second, there are an infinite number of minimizers, but all local minimizers are global minimizers. In the third example, there are many local minimizers that are not global minimizers.

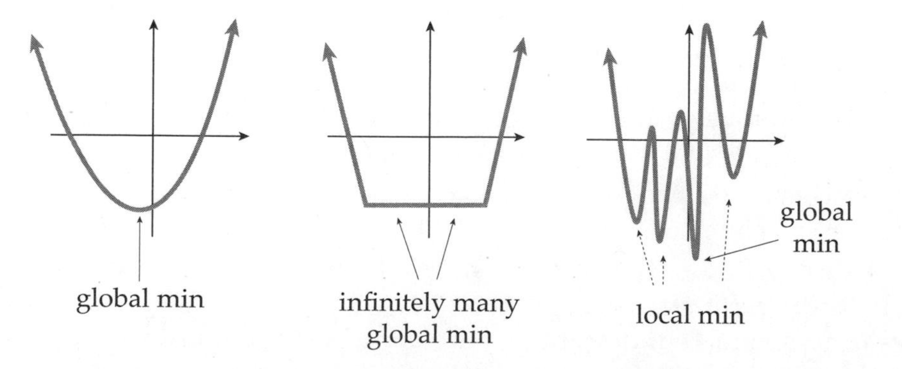

Figure 5.1: Examples of minima of functions

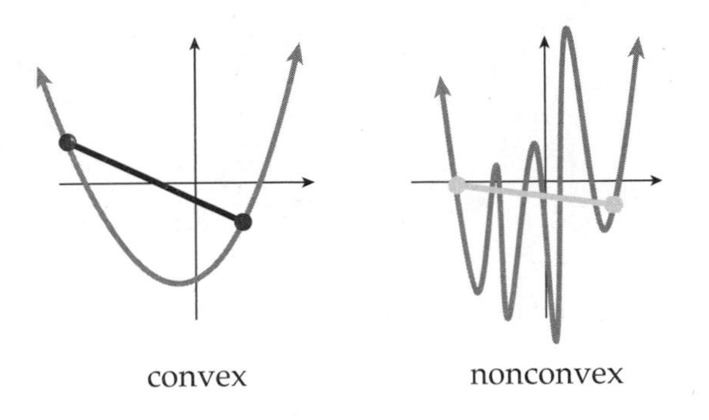

Figure 5.2: Convex versus nonconvex functions

Note that in our example, the two functions without suboptimal local minimizers share the property that for any two points w_1 and w_2, the line segment connecting $(w_1, \Phi(w_1))$ to $(w_2, \Phi(w_2))$ lies completely above the graph of the function. Such functions are called *convex functions*.

Definition 6. *A function Φ is* convex *if for all w_1, w_2 in $\mathbb{R}^d$ and $\alpha \in [0,1]$,*

$$\Phi(\alpha w_1 + (1-\alpha)w_2) \le \alpha\Phi(w_1) + (1-\alpha)\Phi(w_2)\,.$$

We will see shortly that convex functions are the class of functions where gradient descent is guaranteed to find an optimal solution.

Gradient descent

Suppose we want to minimize a differentiable function $\Phi\colon \mathbb{R}^d \to \mathbb{R}$. Most of the algorithms we will consider start at some point w_0 and then aim to find a new point w_1 with a lower function value. The simplest way to do so is to find a direction v such that Φ is decreasing when moving along the direction v. This notion can be formalized by the following definition:

Definition 7. *A vector v is a* descent direction *for Φ at w_0 if $\Phi(w_0 + tv) < \Phi(w_0)$ for some $t > 0$.*

For continuously differentiable functions, it's easy to tell if v is a descent direction: if $v^T\nabla\Phi(w_0) < 0$ then v is a descent direction.

To see this, note that by Taylor's theorem,

$$\Phi(w_0 + \alpha v) = \Phi(w_0) + \alpha\nabla\Phi(w_0 + \tilde{\alpha}v)^T v$$

for some $\tilde{\alpha} \in [0, \alpha]$. By continuity, if α is small, we'll have $\nabla\Phi(w_0 + \tilde{\alpha}v)^T v < 0$. Therefore $\Phi(w_0 + \alpha v) < \Phi(w_0)$ and v is a descent direction.

This characterization of descent directions allows us to provide conditions as to when w minimizes Φ.

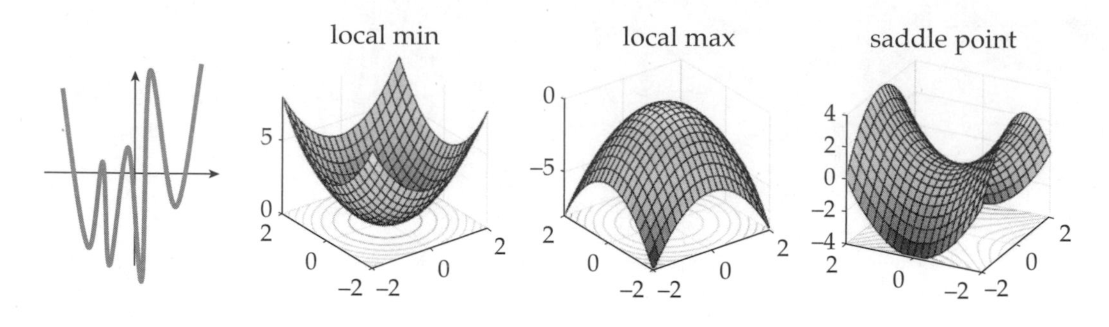

Figure 5.3: Examples of stationary points

Proposition 4. *The point $w_\star$ is a local minimizer only if $\nabla\Phi(w_\star) = 0$.*

Why is this true? Well, the point $-\nabla\Phi(w_\star)$ is always a descent direction if it's not zero. If $w_\star$ is a local minimum, there can be no descent directions. Therefore, the gradient must vanish.

Gradient descent uses the fact that the negative gradient is always a descent direction to construct an algorithm: repeatedly compute the gradient and take a step in the opposite direction to minimize Φ.

Gradient Descent

- Start from an initial point $w_0 \in \mathbb{R}^d$.
- At each step $t = 0, 1, 2, \ldots$:
  - Choose a step size $\alpha_t > 0$.
  - Set $w_{t+1} = w_t - \alpha_t \nabla\Phi(w_t)$.

Gradient descent terminates whenever the gradient is so small that the iterates w_t no longer substantially change. Note now that there can be points where the gradient vanishes but where the function is not minimized. For example, maxima have this property. In general, points where the gradient vanishes are called *stationary points*. It is critically important to remember that not all stationary points are minimizers.

For convex Φ, the situation is dramatically simpler. This is part of the reason why convexity is so appealing.

Proposition 5. *Let $\Phi : \mathbb{R}^d \to \mathbb{R}$ be a differentiable convex function. Then $w_\star$ is a global minimizer of Φ if and only if $\nabla\Phi(w_\star) = 0$.*

Proof. To prove this, we need our definition of convexity: for any $\alpha \in [0, 1]$ and $w \in \mathbb{R}^d$,

$$\Phi(w_\star + \alpha(w - w_\star)) = \Phi((1-\alpha)w_\star + \alpha w) \leq (1-\alpha)\Phi(w_\star) + \alpha\Phi(w) .$$

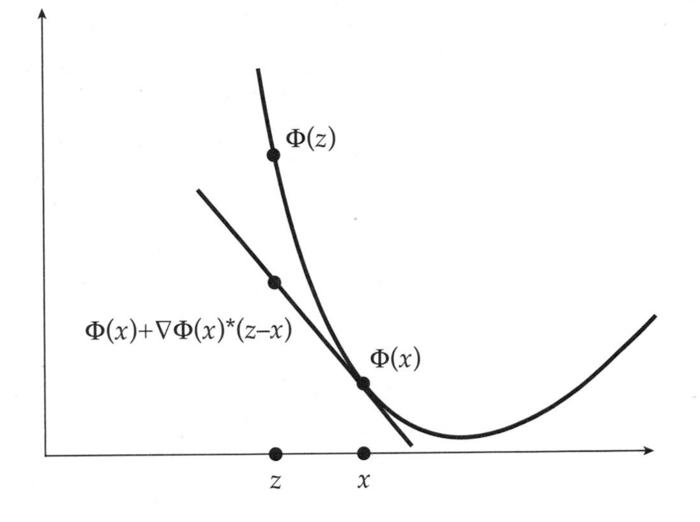

Figure 5.4: Tangent planes to graphs of functions are defined by the gradient.

Here, the inequality is just our definition of convexity. Now, if we rearrange terms, we have

$$\Phi(w) \geq \Phi(w_\star) + \frac{\Phi(w_\star + \alpha(w - w_\star)) - \Phi(w_\star)}{\alpha}.$$

Now apply Taylor's theorem: there is now some $\tilde{\alpha} \in [0,1]$ such that $\Phi(w_\star + \alpha(w - w_\star)) - \Phi(w_\star) = \alpha \nabla\Phi(w_\star + \tilde{\alpha}(w - w_\star))^T(w - w_\star)$. Taking the limit as α goes to zero yields

$$\Phi(w) \geq \Phi(w_\star) + \nabla\Phi(w_\star)^T(w - w_\star).$$

But if $\nabla\Phi(w_\star) = 0$, that means, $\Phi(w) \geq \Phi(w_\star)$ for all w, and hence $w_\star$ is a global minimizer.

□

Tangent hyperplanes always fall below the graphs of convex functions.

Proposition 6. *Let $\Phi : \mathbb{R}^d \to \mathbb{R}$ be a differentiable convex function. Then for any u and v, we have*

$$\Phi(u) \geq \Phi(v) + \nabla\Phi(v)^T(u - v).$$

A convex function cookbook

Testing if a function is convex can be tricky in more than two dimensions. But here are five rules that generate convex functions from simpler functions. In machine learning, almost all convex cost functions are built using these rules.

1. All norms are convex (this follows from the triangle inequality).

2. If Φ is convex and $\alpha \geq 0$, then $\alpha\Phi$ is convex.

3. If Φ and Ψ are convex, then $\Phi + \Psi$ is convex.

4. If Φ and Ψ are convex, then $h(w) = \max\{\Phi(w), \Psi(w)\}$ is convex.

5. If Φ is convex and A is a matrix and b is a vector, then the function $h(w) = \Phi(Aw + b)$ is convex.

All of these properties can be verified using only the definition of convex functions. For example, consider the fourth property. This is probably the trickiest of the list. Take two points w_1 and w_2 and $\alpha \in [0, 1]$. Suppose, without loss of generality, that $\Phi((1-\alpha)w_1 + \alpha w_2) \geq \Psi((1-\alpha)w_1 + \alpha w_2)$

$$\begin{aligned} h((1-\alpha)w_1 + \alpha w_2) &= \max\{\Phi((1-\alpha)w_1 + \alpha w_2), \Psi((1-\alpha)w_1 + \alpha w_2)\} \\ &= \Phi((1-\alpha)w_1 + \alpha w_2) \\ &\leq (1-\alpha)\Phi(w_1) + \alpha\Phi(w_2) \\ &\leq (1-\alpha)\max\{\Phi(w_1), \Psi(w_1)\} + \alpha\max\{\Phi(w_2), \Psi(w_2)\} \\ &= (1-\alpha)h(w_1) + \alpha h(w_2) . \end{aligned}$$

Here, the first inequality follows because Φ is convex. Everything else follows from the definition that h is the max of Φ and Ψ. The reader should verify the other four assertions as an exercise. Another useful exercise is to verify that the SVM cost in the next section is convex by just using these five basic rules and the fact that the one-dimensional function $f(x) = mx + b$ is convex for any scalars m and b.

Applications to empirical risk minimization

For decision theory problems, we studied the zero-one loss that counts errors:

$$loss(\widehat{y}, y) = \mathbb{1}\{y\widehat{y} < 0\} .$$

Unfortunately, this loss is not useful for the gradient method. The gradient is zero almost everywhere. As we discussed in the chapter on supervised learning, machine learning practice always turns to surrogate losses that are easier to optimize. Here we review three popular choices, all of which are convex loss functions. Each choice leads to a different important optimization problem that has been studied in its own right.

The support vector machine

Consider the canonical problem of support vector machine classification. We are provided pairs (x_i, y_i), with $x_i \in \mathbb{R}^d$ and $y_i \in \{-1, 1\}$ for $i = 1, \ldots n$ (note, the y labels are now in $\{-1, 1\}$ instead of $\{0, 1\}$). The goal is to find a vector $w \in \mathbb{R}^d$ such that:

$$\begin{aligned} w^T x_i &> 0 \quad \text{for } y_i = 1 \\ w^T x_i &< 0 \quad \text{for } y_i = -1 . \end{aligned}$$

Such a w defines a half-space where we believe all of the positive examples lie on one side and the negative examples on the other.

Rather than classifying these points exactly, we can allow some slack. We can pay a penalty of $1 - y_i w^T x_i$ points that are not strongly classified. This motivates the hinge loss we encountered earlier and leads to the *support vector machine objective*:

$$\text{minimize}_w \quad \sum_{i=1}^{n} \max\left\{1 - y_i w^T x_i, 0\right\} .$$

Defining the function $e(z) = \mathbb{1}\{z \leq 1\}$, we can compute that the gradient of the SVM cost is

$$-\sum_{i=1}^{n} e(y_i w^T x_i) y_i x_i .$$

Hence, gradient descent for this ERM problem would follow the iteration

$$w_{t+1} = w_t + \alpha \sum_{i=1}^{n} e(y_i w^T x_i) y_i x_i .$$

Although similar, note that this isn't quite the perceptron method yet. The time to compute one gradient step is $O(n)$ as we sum over all n inner products. We will soon turn to the stochastic gradient method that has constant iteration complexity and will subsume the perceptron algorithm.

Logistic regression

Logistic regression is equivalent to using the loss function

$$loss(\widehat{y}, y) = \log\left(1 + \exp(-y\widehat{y})\right) .$$

Note that even though this loss has a probabilistic interpretation, it can also just be seen as an approximation to the error-counting zero-one loss.

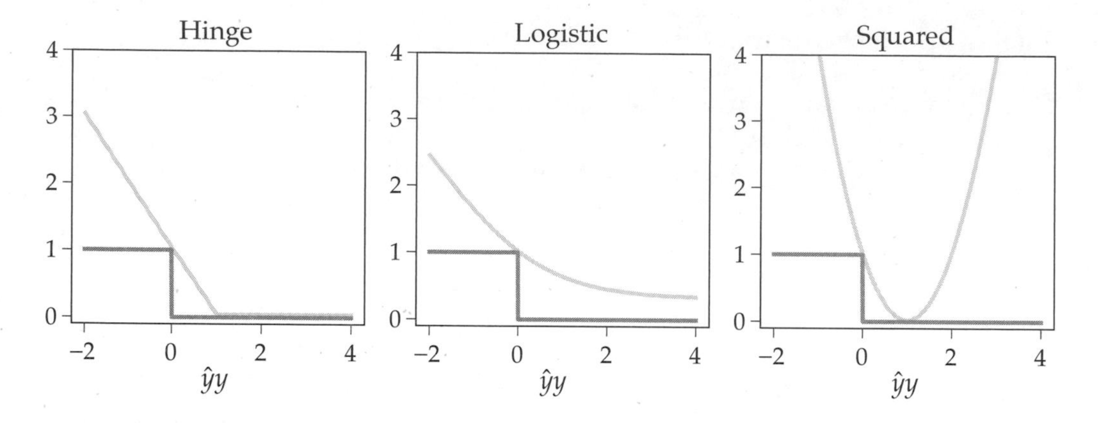

Figure 5.5: Three different convex losses compared with the zero-one loss

Least squares classification

Least squares classification uses the loss function

$$loss(\widehat{y}, y) = \tfrac{1}{2}(\widehat{y} - y)^2 .$$

Though this might seem like an odd approximation to the error-counting loss, it leads to the maximum a posteriori (MAP) decision rule when minimizing the population risk. Recall the MAP rule selects the label that has the highest probability conditional on the observed data.

It is helpful to keep in mind the next picture, which summarizes how each of these different loss functions approximate the zero-one loss. We can ensure that the squared loss is an upper bound on the zero-one loss by dropping the factor 1/2.

Insights from quadratic functions

Quadratic functions are the prototypical example that motivate algorithms for differentiable optimization problems. Though not all insights from quadratic optimization transfer to more general functions, there are several key features of the dynamics of iterative algorithms on quadratics that are notable. Moreover, quadratics are a good test case for reasoning about optimization algorithms: if a method doesn't work well on quadratics, it typically won't work well on more complicated nonlinear optimization problems. Finally, note that ERM with linear functions and a squared loss is a quadratic optimization problem, so such problems are indeed relevant to machine learning practice.

The general quadratic optimization problem takes the form

$$\Phi(w) = \tfrac{1}{2}w^T Q w - p^T w + r ,$$

where Q is a symmetric matrix, p is a vector, and r is a scalar. The scalar r only affects the value of the function, and plays no role in the dynamics of gradient descent. The gradient of this function is

$$\nabla\Phi(w) = Qw - p\,.$$

The stationary points of Φ are the w where $Qw = p$. If Q is full rank, there is a unique stationary point.

The gradient descent algorithm for quadratics follows the iterations

$$w_{t+1} = w_t - \alpha(Qw_t - p)\,.$$

If we let $w_\star$ be *any* stationary point of Φ, we can rewrite this iteration as

$$w_{t+1} - w_\star = (I - \alpha Q)(w_t - w_\star)\,.$$

Unwinding the recursion yields the "closed form" formula for the gradient descent iterates:

$$w_t - w_\star = (I - \alpha Q)^t(w_0 - w_\star)\,.$$

This expression reveals several possible outcomes. Let $\lambda_1 \geq \lambda_2 \geq \ldots \geq \lambda_d$ denote the eigenvalues of Q. These eigenvalues are real because Q is symmetric. First suppose that Q has a negative eigenvalue $\lambda_d < 0$ and v is an eigenvector such that $Qv = \lambda_d v$. Then $(I - \alpha Q)^t v = (1 + \alpha|\lambda_d|)^t v$ which tends to infinity as t grows. This is because $1 + \alpha|\lambda_d|$ is greater than 1 if $\alpha > 0$. Hence, if $\langle v, w_0 - w_\star\rangle \neq 0$, gradient descent *diverges*. For a random initial condition w_0, we'd expect this dot product will not equal zero, and hence gradient descent will almost surely not converge from a random initialization.

In the case that all of the eigenvalues of Q are positive, choosing α greater than zero and less than $1/\lambda_1$ will ensure that $0 \leq 1 - \alpha\lambda_k < 1$ for all k. In this case, the gradient method converges exponentially quickly to the optimum $w_\star$:

$$\begin{aligned}\|w_{t+1} - w_\star\| &= \|(I - \alpha Q)(w_t - w_\star)\| \\ &\leq \|I - \alpha Q\|\|w_t - w_\star\| \leq \left(1 - \frac{\lambda_d}{\lambda_1}\right)\|w_t - w_\star\|\,.\end{aligned}$$

When the eigenvalues of Q are all positive, the function Φ is strongly convex. Strongly convex functions turn out to be the set of functions where gradient descent with a constant step size converges exponentially from any starting point.

Note that the ratio of λ_1 to λ_d governs how quickly all of the components converge to 0. Defining the *condition number* of Q to be $\kappa = \lambda_1/\lambda_d$ and setting the step size $\alpha = 1/\lambda_1$, gives the bound

$$\|w_t - w_\star\| \leq \left(1 - \kappa^{-1}\right)^t\|w_0 - w_\star\|\,.$$

This rate reflects what happens in practice: when there are small singular values, gradient descent tends to bounce around and oscillate as shown in the figure below. When the condition number of Q is small, gradient descent makes rapid progress toward the optimum.

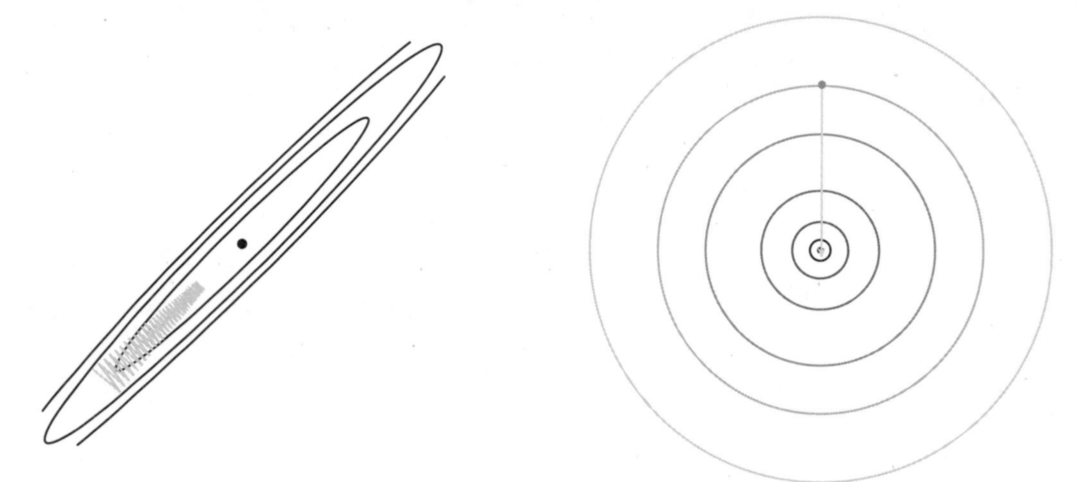

Figure 5.6: Oscillating behavior versus rapid convergence

There is one final case that's worth considering. When all of the eigenvalues of Q are nonnegative but some of them are zero, the function Φ is convex but not strongly convex. In this case, exponential convergence to a unique point cannot be guaranteed. In particular, there will be an infinite number of global minimizers of Φ. If $w_\star$ is a global minimizer and v is any vector with $Qv = 0$, then $w_\star + v$ is also a global minimizer. However, in the span of the eigenvectors corresponding to positive eigenvalues, gradient descent still converges exponentially. For general convex functions, it will be important to consider different parts of the parameter space to fully understand the dynamics of gradient methods.

Stochastic gradient descent

The stochastic gradient method is one of the most popular algorithms for contemporary data analysis and machine learning. It has a long history and has been "invented" several times by many different communities (under the names "least mean squares," "backpropagation," "online learning," and the "randomized Kaczmarz method"). Most researchers attribute this algorithm to the initial work of Robbins and Monro from 1951 who solved a more general problem with the same method.[76]

Consider again our main goal of minimizing the empirical risk with respect to a vector of parameters w, and consider the simple case of linear classification where w is d-dimensional and

$$f(x_i; w) = w^T x_i .$$

The idea behind the stochastic gradient method is that since the gradient of a sum is the sum of the gradients of the summands, each summand provides useful information about how to optimize the total sum. Stochastic gradient descent minimizes empirical risk by following the gradient of the risk evaluated on a *single, random* example.

Stochastic Gradient Descent

- Start from an initial point $w_0 \in \mathbb{R}^d$.
- At each step $t = 0, 1, 2, \ldots$:
  - Choose a step size $\alpha_t > 0$ and random index $i \in [n]$.
  - Set $w_{t+1} = w_t - \alpha_t \nabla_{w_t} loss(f(x_i; w_t), y_i)$.

The intuition behind this method is that by following a descent direction in expectation, we should be able to get close to the optimal solution if we wait long enough. However, it's not quite that simple. Note that even when the gradient of the sum is zero, the gradients of the individual summands may not be. The fact that $w_\star$ is no longer a fixed point complicates the analysis of the method.

Example: Revisiting the perceptron

Let's apply the stochastic gradient method to the support vector machine cost loss. We initialize our half-space at some w_0. At iteration t, we choose a random data point (x_i, y_i) and update

$$w_{t+1} = w_t + \eta \begin{cases} y_i x_i & \text{if } y_i w_t^T x_i \leq 1 \\ 0 & \text{otherwise} . \end{cases}$$

As we promised earlier, we see that using stochastic gradient descent to minimize empirical risk with a hinge loss is completely equivalent to Rosenblatt's perceptron algorithm.

Example: Computing a mean

Let's now try to examine the simplest example possible. Consider applying the stochastic gradient method to the function

$$\frac{1}{2n} \sum_{i=1}^{n} (w - y_i)^2 ,$$

where $y_1, \ldots, y_n$ are fixed scalars. This setup corresponds to a rather simple classification problem where the x features are all equal to 1. Note that the gradient of one of the increments is

$$\nabla loss(f(x_i; w), y) = w - y_i .$$

To simplify notation, let's imagine that our random samples are coming to us in order $\{1, 2, 3, 4, ...\}$. Start with $w_1 = 0$, use the step size $\alpha_k = 1/k$. We can then write out the first few equations:

$$\begin{aligned} w_2 &= w_1 - w_1 + y_1 = y_1 \\ w_3 &= w_2 - \frac{1}{2}(w_2 - y_2) = \frac{1}{2}y_1 + \frac{1}{2}y_2 \\ w_4 &= w_3 - \frac{1}{3}(w_3 - y_3) = \frac{1}{3}y_1 + \frac{1}{3}y_2 + \frac{1}{3}y_3 . \end{aligned}$$

Generalizing from here, we can conclude by induction that

$$w_{k+1} = \left(\frac{k-1}{k}\right) w_k + \frac{1}{k} y_k = \frac{1}{k} \sum_{i=1}^{k} y_i .$$

After n steps, w_n is the mean of the y_i, and you can check by taking a gradient that this is indeed the minimizer of the ERM problem.

The $1/k$ step size was the originally proposed step size by Robbins and Monro. This simple example justifies why: we can think of the stochastic gradient method as computing a running average. Another motivation for the $1/k$ step size is that the steps tend to zero, but the path length is infinite.

Moving to a more realistic random setting where the data might arrive in any order, consider what happens when we run the stochastic gradient method on the function

$$R(w) = \tfrac{1}{2}\mathbb{E}[(w - Y)^2] .$$

Here Y is some random variable with mean μ and variance σ^2. If we run for k steps with i.i.d. samples Y_i at each iteration, the calculation above reveals that

$$w_k = \frac{1}{k} \sum_{i=1}^{k} Y_i .$$

The associated cost is

$$R(w_k) = \frac{1}{2} \mathbb{E}\left[\left(\frac{1}{k}\sum_{i=1}^{k} Y_i - Y\right)^2\right] = \frac{1}{2k}\sigma^2 + \frac{1}{2}\sigma^2 .$$

Compare this to the minimum achievable risk $R_\star$. Expanding the definition

$$R(w) = \tfrac{1}{2}\mathbb{E}[w^2 - 2Yw + Y^2] = \tfrac{1}{2}w^2 - \mu w + \tfrac{1}{2}\sigma^2 + \tfrac{1}{2}\mu^2 ,$$

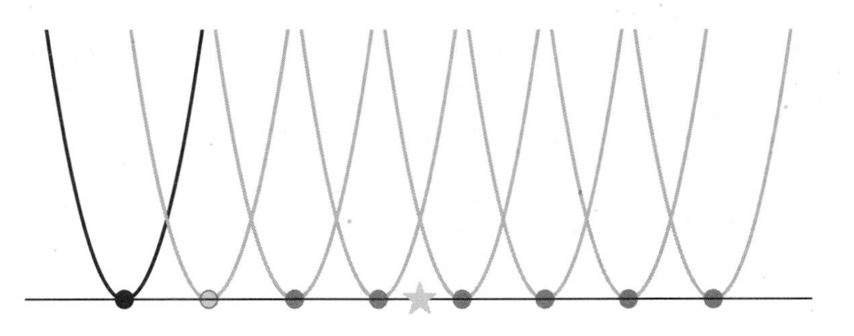

Figure 5.7: Plot of the different increments of $\frac{1}{2n}\sum_{i=1}^{n}(w - y_i)^2$. The star denotes the optimal solution.

we find that the minimizer is $w_\star = \mu$. Its cost is

$$R_\star = R(w_\star) = \frac{1}{2}\sigma^2\,,$$

and after n iterations, we have the expected *optimality gap*

$$\mathbb{E}\left[R(w_n) - R_\star\right] = \frac{1}{2n}\sigma^2\,.$$

This is the best we could have achieved using any estimator for $w_\star$ given the collection of random draws. Interestingly, the incremental "one-at-a-time" method finds as good a solution as one that considers all of the data together. This basic example reveals a fundamental limitation of the stochastic gradient method: we can't expect to generically get fast convergence rates without additional assumptions. Statistical fluctuations themselves prevent the optimality gap from decreasing exponentially quickly.

This simple example also helps give intuition on the convergence as we sample stochastic gradients. The figure below plots an example of each individual term in the summand, shaded to distinguish each term. The minimizing solution is marked with a star. To the far left and far right of the figure, all of the summands will have gradients pointing in the same direction to the solution. However, as our iterate gets close to the optimum, we will be pointed in different directions depending on which gradient we sample. By reducing the step size, we will be more likely to stay close and eventually converge to the optimal solution.

Example: Stochastic minimization of a quadratic function

Let's consider a more general version of stochastic gradient descent that follows the gradient plus some unbiased noise. We can model the algorithm as minimizing a function $\Phi(w)$ where we follow a direction $\nabla\Phi(w) + \nu$ where ν

is a random vector with mean zero and $\mathbb{E}[\|\nu\|^2] = \sigma^2$. Consider the special case where $\Phi(w)$ is a quadratic function

$$\Phi(w) = \tfrac{1}{2} w^T Q w - p^T w + r \,.$$

Then the iterations take the form

$$w_{t+1} = w_t - \alpha(Q w_t - p + \nu_t) \,.$$

Let's consider what happens when Q is positive definite with maximum eigenvalue λ_1 and minimum eigenvalue $\lambda_d > 0$. Then, if $w_\star$ is a global minimizer of Φ, we can again rewrite the iterations as

$$w_{t+1} - w_\star = (I - \alpha Q)(w_t - w_\star) - \alpha \nu_t \,.$$

Since we assume that the noise ν_t is independent of all of the w_k with $k \leq t$, we have

$$\mathbb{E}[\|w_{t+1} - w_\star\|^2] \leq \|I - \alpha Q\|^2 \, \mathbb{E}[\|w_t - w_\star\|^2] + \alpha^2 \sigma^2 \,,$$

which looks like the formula we derived for quadratic functions, but now with an additional term from the noise. Assuming $\alpha < 1/\lambda_1$, we can unwind this recursion to find

$$\mathbb{E}[\|w_t - w_\star\|^2] \leq (1 - \alpha \lambda_d)^{2t} \|w_0 - w_\star\|^2 + \frac{\alpha \sigma^2}{\lambda_d} \,.$$

From this expression, we see that gradient descent converges exponentially quickly to some ball around the optimal solution. The smaller we make α, the closer we converge to the optimal solution, but the rate of convergence is also slower for smaller α. This trade-off motivates many of the step size selection rules in stochastic gradient descent. In particular, it is common to start with a large step size and then successively reduce the step size as the algorithm progresses.

Tricks of the trade

In this section, we describe key engineering techniques that are useful for tuning the performance of stochastic gradient descent. Every machine learning practitioner should know these simple tricks.

Shuffling. Even though we described the stochastic gradient method as a sampling each gradient with replacement from the increments, in practice better results are achieved by simply randomly permuting the data points and then running SGD in this random order. This is called "shuffling," and even a single shuffle can eliminate the pathological behavior we described in the example with highly correlated data. Recently, beginning with work by Gürbüzbalaban et al., researchers have shown that in theory, shuffling outperforms independent sampling of increments.[77] The arguments for without-replacement sampling remain more complicated than those for with-replacement derivations, but optimal sampling for SGD remains an active area of optimization research.

Step size selection. Step size selection in SGD remains a hotly debated topic. We saw above a decreasing step size $1/k$ solved our simple one-dimensional ERM problem. However, a rule that works for an unreasonable number of cases is to simply pick the largest step size that does not result in divergence. This step will result in a model that is not necessarily optimal, but significantly better than initialization. By slowly reducing the step size from this initial large step size to successively smaller ones, we can zero in on the optimal solution.

Step decay. The step size is usually reduced after a fixed number of passes over the training data. A pass over the entire dataset is called an *epoch*. In an epoch, some number of iterations are run, and then a choice is made about whether to change the step size. A common strategy is to run with a constant step size for some fixed number of iterations T, and then reduce the step size by a constant factor γ. Thus, if our initial step size is α, on the kth epoch, the step size is $\alpha\gamma^{k-1}$. This method is often more robust in practice than the diminishing step size rule. For this step size rule, a reasonable heuristic is to choose γ between 0.8 and 0.9. Sometimes people choose rules as aggressive as $\gamma = 0.1$.

Another possible schedule for the step size is called *epoch doubling*. In epoch doubling, we run for T steps with step size α, then run $2T$ steps with step size $\alpha/2$, and then $4T$ steps with step size $\alpha/4$, and so on. Note that this provides a piecewise constant approximation to the function α/k.

Minibatching. A common technique used to take advantage of parallelism is called *minibatching*. A minibatch is an average of many stochastic gradients. Suppose at each iteration we sample a batch$_k$ with m data points. The update rule then becomes

$$w_{k+1} = w_k - \alpha_k \frac{1}{m} \sum_{j \in \text{batch}_k} \nabla_w loss(f(x_j; w_k), y_j) .$$

Minibatching reduces the variance of the stochastic gradient estimate of the true gradient, and hence tends to be a better descent direction. Of course, there are trade-offs in total computation time versus the size of the minibatch, and these typically need to be handled on a case by case basis.

Momentum. Finally, we note that one can run stochastic gradient descent with *momentum*. Momentum mixes the current gradient direction with the previously taken step. The idea here is that if the previous weight update was good, we may want to continue moving along this direction. The algorithm iterates are defined as

$$w_{k+1} = w_k - \alpha_k g_k(w_k) + \beta(w_k + w_{k-1}) ,$$

where g_k denotes a stochastic gradient. In practice, these methods are very successful. Typical choices for β here are between 0.8 and 0.95. Momentum can provide significant accelerations, and should be considered an option in any implementation of SGM.

The SGD quick start guide

Newcomers to stochastic gradient descent often find all of these design choices daunting, and it's useful to have simple rules of thumb to get going. We recommend the following:

1. Pick as large a minibatch size as you can given your computer's RAM.
2. Set your momentum parameter to either 0 or 0.9. Your call!
3. Find the largest constant step size such that SGD doesn't diverge. This takes some trial and error, but you only need to be accurate to within a factor of 10 here.
4. Run SGD with this constant step size until the empirical risk plateaus.
5. Reduce the step size by a constant factor (say, 10)
6. Repeat steps 4 and 5 until you converge.

While this approach may not be the most optimal in all cases, it's a great starting point and is good enough for probably 90% of applications we've encountered.

Analysis of the stochastic gradient method

We now turn to a theoretical analysis of the general stochastic gradient method. Before we proceed, let's set up some conventions. We will assume that we are trying to minimize a convex function $R : \mathbb{R}^d \to \mathbb{R}$. Let $w_\star$ denote any optimal solution of R. We will assume we gain access at every iteration to a *stochastic function* $g(w;\xi)$ such that

$$\mathbb{E}_\xi[g(w;\xi)] = \nabla R(w)\,.$$

Here ξ is a random variable that determines what our direction looks like. We additionally assume that these stochastic gradients are *bounded* so there exists a nonnegative constant B such that

$$\|g(w;\xi)\| \le B\,.$$

We will study the stochastic gradient iteration

$$w_{t+1} = w_t - \alpha_t g\,(w_t;\xi_t)\;.$$

Throughout, we will assume that the sequence $\{\xi_j\}$ is selected i.i.d. from some fixed distribution.

We begin by expanding the distance to the optimal solution:

$$\begin{aligned}\|w_{t+1} - w_\star\|^2 &= \|w_t - \alpha_t g_t(w_t; \xi_t) - w_\star\|^2 \\ &= \|w_t - w_\star\|^2 - 2\alpha_t\langle g_t(w_t;\xi_t), w_t - w_\star\rangle + \alpha_t^2\|g_t(w_t;\xi_t)\|^2 .\end{aligned}$$

We deal with each term in this expansion separately. First note that if we apply the law of iterated expectation

$$\begin{aligned}\mathbb{E}[\langle g_t(w_t;\xi_t), w_t - w_\star\rangle] &= \mathbb{E}\left[\mathbb{E}_{\xi_t}[\langle g_t(w_t;\xi_t), w_t - w_\star\rangle \mid \xi_0,\ldots,\xi_{t-1}]\right] \\ &= \mathbb{E}\left[\langle \mathbb{E}_{\xi_t}[g_t(w_t;\xi_t) \mid \xi_0,\ldots,\xi_{t-1}], w_t - w_\star\rangle\right] \\ &= \mathbb{E}\left[\langle \nabla R(w_t), w_t - w_\star\rangle\right] .\end{aligned}$$

Here, we are simply using the fact that ξ_t being independent of all of the preceding ξ_i implies that it is independent of w_t. This means that when we iterate the expectation, the stochastic gradient can be replaced by the gradient. The last term we can bound using our assumption that the gradients are bounded:

$$\mathbb{E}[\|g(w_t;\xi_t)\|^2] \leq B^2 .$$

Letting $\delta_t := \mathbb{E}[\|w_t - w_\star\|^2]$, this gives

$$\delta_{t+1} \leq \delta_t - 2\alpha_t \mathbb{E}\left[\langle \nabla R(w_t), w_t - w_\star\rangle\right] + \alpha_t^2 B^2 .$$

Now let $\lambda_t = \sum_{j=0}^t \alpha_j$ denote the sum of all the step sizes up to iteration t. Also define the average of the iterates weighted by the step size

$$\bar{w}_t = \lambda_t^{-1}\sum_{j=0}^t \alpha_j w_j .$$

We are going to analyze the deviation of $R(\bar{w}_t)$ from optimality.

Also let $\rho_0 = \|w_0 - w_\star\|$. ρ_0 is the initial distance to an optimal solution. It is not necessarily a random variable.

To proceed, we just expand the following expression:

$$\begin{aligned}\mathbb{E}\left[R(\bar{w}_T) - R(w_\star)\right] &\leq \mathbb{E}\left[\lambda_T^{-1}\sum_{t=0}^T \alpha_t(R(w_t) - R(w_\star))\right] \\ &\leq \lambda_T^{-1}\sum_{t=0}^T \alpha_t \mathbb{E}[\langle \nabla R(x_t), w_t - w_\star\rangle] \\ &\leq \lambda_T^{-1}\sum_{t=0}^T \tfrac{1}{2}(\delta_t - \delta_{t+1}) + \tfrac{1}{2}\alpha_t^2 B^2 \\ &= \frac{\delta_0 - \delta_{T+1} + B^2\sum_{t=0}^T \alpha_t^2}{2\lambda_T} \\ &\leq \frac{\rho_0^2 + B^2\sum_{t=0}^T \alpha_t^2}{2\sum_{t=0}^T \alpha_t} .\end{aligned}$$

Here, the first inequality follows because R is convex (the line segments lie above the function, i.e., $R(w_\star) \geq R(w_t) + \langle \nabla R(w_t), w_t - w_\star \rangle$). The second inequality uses the fact that gradients define tangent planes to R and always lie below the graph of R, and the third inequality uses the recursion we derived above for δ_t.

The analysis we saw gives the following result.[78]

Theorem 3. *Suppose we run the SGM on a convex function R with minimum value $R_\star$ for T steps with step size α. Define*

$$\alpha_{\text{opt}} = \frac{\rho_0}{B\sqrt{T}} \qquad \textit{and} \qquad \theta = \frac{\alpha}{\alpha_{\text{opt}}}.$$

Then, we have the bound

$$\mathbb{E}[R(\bar{w}_T) - R_\star] \leq \left(\tfrac{1}{2}\theta + \tfrac{1}{2}\theta^{-1}\right) \frac{B\rho_0}{\sqrt{T}}.$$

This proposition asserts that we pay linearly for errors in selecting the optimal constant step size. If we guess a constant step size that is two times or one half of the optimal choice, then we need to run for at most twice as many iterations. The optimal step size is found by minimizing our upper bound on the suboptimality gap. Other step sizes could also be selected here, including diminishing step size. But the constant step size turns out to be optimal for this upper bound.

What are the consequences for risk minimization? First, for *empirical risk*, assume we are minimizing a convex loss function and searching for a linear predictor. Assume further that there exists a model with zero empirical risk. Let C be the maximum value of the gradient of the loss function, D be the largest norm of any example x_i, and let ρ denote the minimum norm w such that $R_S[w] = 0$. Then we have

$$\mathbb{E}[R_S[\bar{w}_T]] \leq \frac{CD\rho}{\sqrt{T}}.$$

We see that with appropriately chosen step size, the stochastic gradient method converges at a rate of $1/\sqrt{T}$, the same rate of convergence observed when studying the one-dimensional mean computation problem. Again, the stochasticity forces us into a slow $1/\sqrt{T}$ rate of convergence, but high dimensionality does not change this rate.

Second, if we only operate on each sample exactly once, and we assume our data is i.i.d., we can think of the stochastic gradient method as minimizing the *population risk* instead of the empirical risk. With the same notation, we'll have

$$\mathbb{E}[R[\bar{w}_T]] - R_\star \leq \frac{CD\rho}{\sqrt{T}}.$$

The analysis of stochastic gradient gives our second *generalization bound* of the book. What it shows is that by optimizing over a fixed set of T data points, we can get a solution that will have low cost on new data. We will return to this observation in the next chapter.

Implicit convexity

We have thus far focused on convex optimization problems, showing that gradient descent can find global minima with modest computational means. What about nonconvex problems? Nonconvex optimization is such a general class of problems that in general it is hard to make useful guarantees. However, ERM is a special optimization problem, and its structure enables nonconvexity to enter in a graceful way.

As it turns out, there's a "hidden convexity" of ERM problems that shows that the *predictions* should converge to a global optimum even if we can't analyze where exactly the model converges. We will show that this insight has useful benefits when models are overparameterized or nonconvex.

Suppose we have a loss function *loss* that is equal to zero when $\widehat{y} = y$ and is nonnegative otherwise. Suppose we have a generally parameterized function class $\{f(x;w) \colon w \in \mathbb{R}^d\}$ and we aim to find parameters w that minimize the empirical risk. The empirical risk

$$R_S[w] = \frac{1}{n}\sum_{i=1}^{n} loss(f(x_i;w), y_i)$$

is bounded below by 0. Hence if we can find a solution with $f(x_i;w) = y_i$ for all i, we would have a *global minimum*, not a local minimum. This is a trivial observation, but one that helps focus our study. If during optimization all predictions $f(x_i;w)$ converge to y_i for all i, we will have computed a global minimizer. For the sake of simplicity, we specialize to the square loss in this section:

$$loss(f(x_i;w), y_i) = \tfrac{1}{2}(f(x_i;w) - y_i)^2$$

The argument we develop here is inspired by the work of Du et al. who use a similar approach to rigorously analyze the convergence of two layer neural networks.[79] Similar calculations can be made for other losses with some modifications of the argument.

Convergence of overparameterized linear models

Let's first consider the case of linear prediction functions

$$f(x;w) = w^T x\,.$$

Define

$$y = \begin{bmatrix} y_1 \\ \vdots \\ y_n \end{bmatrix} \qquad \text{and} \qquad X = \begin{bmatrix} x_1^T \\ \vdots \\ x_n^T \end{bmatrix}.$$

We can then write the empirical risk objective as

$$R_S[w] = \frac{1}{2n}\|Xw - y\|^2.$$

The gradient descent update rule has the form

$$w_{t+1} = w_t - \alpha X^T(Xw_t - y).$$

We pull the scaling factor $1/n$ into the step size for notational convenience. Now define the vector of predictions

$$\hat{y}_t = \begin{bmatrix} f(x_1; w_t) \\ \vdots \\ f(x_n; w_t) \end{bmatrix}.$$

For the linear case, the predictions are given by $\hat{y}_k = Xw_k$. We can use this definition to track the evolution of the *predictions* instead of the parameters w. The predictions evolve according to the rule

$$\hat{y}_{t+1} = \hat{y}_t - \alpha XX^T(\hat{y}_t - y).$$

This looks a lot like the gradient descent algorithm applied to a strongly convex quadratic function that we studied earlier. Subtracting y from both sides and rearranging shows

$$\hat{y}_{t+1} - y = (I - \alpha XX^T)(\hat{y}_t - y).$$

This expression proves that as long as XX^T is strictly positive definite and α is small enough, the predictions converge to the training labels. Keep in mind that X is $n \times d$ and a model is overparameterized if $d > n$. The $n \times n$ matrix XX^T has a chance of being strictly positive definite in this case.

When we use a sufficiently small and constant step size α, our predictions converge at an *exponential* rate. This is in contrast to the behavior we saw for gradient methods on overdetermined problems. Our general analysis of the weights showed that the convergence rate might be only inverse polynomial in the iteration counter. In the overparameterized regime, we can guarantee the predictions converge more rapidly than the weights themselves.

The rate in this case is governed by properties of the matrix X. As we have discussed, we need the eigenvalues of XX^T to be positive, and we'd ideally like that the eigenvalues be all of similar magnitude.

First note that a *necessary* condition is that d, the dimension, must be larger than n, the number of data points. That is, we need to have an overparameterized model in order to ensure exponentially fast convergence of the predictions. We have already seen that such overparameterized models make it possible to interpolate any set of labels and to always force the data to be linearly separable. Here, we see further that overparameterization encourages optimization methods to converge in fewer iterations by improving the condition number of the data matrix.

Overparameterization can also help accelerate convergence. Recall that the eigenvalues of XX^T are the squares of the *singular values* of X. Let us write out a singular value decomposition $X = USV^T$, where S is a diagonal matrix of singular values $(\sigma_1, \ldots, \sigma_n)$. In order to improve the condition number of this matrix, it suffices to add a feature that is concentrated in the span of the singular vectors with small singular values. How to find such features is not always apparent, but does give us a starting point as to where to look for new, informative features.

Convergence of nonconvex models

Surprisingly, this analysis naturally extends to nonconvex models. With some abuse of notation, let $\hat{y} = f(x; w) \in \mathbb{R}^n$ denote the n predictions of some nonlinear model parameterized by the weights w on input x. Our goal is to minimize the squared loss objective

$$\frac{1}{2}\|f(x; w) - y\|^2 .$$

Since the model is nonlinear this objective is no longer convex. Nonetheless we can mimic the analysis we did previously for overparameterized linear models.

Running gradient descent on the weights gives

$$w_{t+1} = w_t - \alpha \mathsf{D}f(x; w_t)(\hat{y}_t - y) ,$$

where $\hat{y}_t = f(x; w_t)$ and $\mathsf{D}f$ is the Jacobian of the predictions with respect to w. That is, $\mathsf{D}f(x; w)$ is the $d \times n$ matrix of first order derivatives of the function $f(x; w)$ with respect to w. We can similarly define the Hessian operator $H(w)$ to be the $n \times d \times d$ array of second derivatives of $f(x; w)$. We can think of $H(w)$ as a quadratic form that maps pairs of vectors $(u, v) \in \mathbb{R}^{d \times d}$ to $\mathbb{R}^n$. With these higher order derivatives, Taylor's theorem asserts

$$\begin{aligned}\hat{y}_{t+1} &= f(x, w_{t+1}) \\ &= f(x, w_t) + \mathsf{D}f(x; w_t)^T (w_{t+1} - w_t) \\ &\quad + \int_0^1 H(w_t + s(w_{t+1} - w_t))(w_{t+1} - w_t, w_{t+1} - w_t) \mathrm{d}s .\end{aligned}$$

Since w_t are the iterates of gradient descent, this means that we can write the prediction as

$$\widehat{y}_{t+1} = \widehat{y}_t - \alpha \mathsf{D}f(x; w_t)^T \mathsf{D}f(x; w_t)(\widehat{y}_t - y) + \alpha\epsilon_t\,,$$

where

$$\epsilon_t = \alpha \int_0^1 H(w_t + s(w_{t+1} - w_t))\,(\mathsf{D}f(x; w_t)(\widehat{y}_t - y), \mathsf{D}f(x; w_t)(\widehat{y}_t - y))\,\mathrm{d}s\,.$$

Subtracting the labels y from both sides and rearranging terms gives the recursion

$$\widehat{y}_{t+1} - y = (I - \alpha \mathsf{D}f(x; w_t)^T \mathsf{D}f(x; w_t))(\widehat{y}_t - y) + \alpha\epsilon_t\,.$$

If ϵ_t vanishes, this shows that the predictions again converge to the training labels as long as the eigenvalues of $\mathsf{D}f(x; w_t)^T \mathsf{D}f(x; w_t)$ are strictly positive. When the error vector ϵ_t is sufficiently small, similar dynamics will occur. We expect ϵ_t to not be too large because it is quadratic in the distance from y_t to y and because it is multiplied by the step size α, which can be chosen to be small.

The nonconvexity isn't particularly disruptive here. We just need to make sure our Jacobians have full rank most of the time and that our steps aren't too large. Again, if the number of parameters is larger than the number of data points, then these Jacobians are likely to be positive definite as long as we've engineered them well. But how exactly can we guarantee that our Jacobians are well behaved? We can derive some reasonable ground rules by unpacking how we compute gradients of compositions of functions. More on this follows in our chapter on deep learning.

Regularization

One complication with optimization in the overparameterized regime is that there is an *infinite collection* of models that achieve zero empirical risk. How do we break ties between these and which set of weights should we prefer?

To answer this question we need to take a step back and remember that the goal of supervised learning is not just to achieve zero training error. We also care about performance on data *outside* the training set, and having zero loss on its own doesn't tell us anything about data outside the training set.

As a toy example, imagine we have two sets of data X_{train} and X_{test} where X_{train} has shape $n \times d$ and X_{test} is $m \times d$. Let y_{train} be the training labels and let q be an m-dimensional vector of random labels. Then if $d > (m + n)$ we can find weights w such that

$$\begin{bmatrix} X_{\text{train}} \\ X_{\text{test}} \end{bmatrix} w = \begin{bmatrix} y_{\text{train}} \\ q \end{bmatrix}.$$

That is, these weights would produce zero error on the training set, but error no better than that of random guessing on the testing set. That's not desired behavior! Of course this example is pathological, because in reality we would have no reason to fit random labels against the test set when we create our model.

The main challenge in supervised learning is to design models that achieve low training error while performing well on new data. The main tool used in such problems is called *regularization*. Regularization is the general term for taking a problem with infinitely many solutions and biasing its solution toward a smaller subset of solution space. This is a highly encompassing notion that belongs somewhere between optimization and generalization, and we will need to examine regularization in both chapters.

Sometimes regularization is *explicit* insofar as we have a desired property of the solution in mind that we exercise as a constraint on our optimization problem. Sometimes regularization is *implicit* insofar as algorithm design choices lead to a unique solution, although the properties of this solution might not be immediately apparent.

Here, we take an unconventional tack of working from implicit to explicit, starting with stochastic gradient descent.

Implicit regularization by optimization

Consider again the linear case of gradient descent or stochastic gradient descent

$$w_{t+1} = w_t - \alpha e_t x_t \,,$$

where e_t is the gradient of the loss at the current prediction. Note that if we initialize $w_0 = 0$, then w_t is always in the span of the data. This can be seen by simple induction. This already shows that even though general weights lie in a high dimensional space, SGD searches over a space with dimension at most n, the number of data points.

Now suppose we have a nonnegative loss with $\frac{\partial loss(z,y)}{\partial z} = 0$ if and only if $y = z$. This condition is satisfied by the square loss, but not hinge and logistic losses. For such losses, at optimality we have for some vector v that:

1. $Xw = y$, because we have zero loss.
2. $w = X^T v$, because we are in the span of the data.

Under the mild assumption that our examples are linearly independent, we can combine these equations to find that

$$w = X^T (XX^T)^{-1} y \,.$$

That is, when we run stochastic gradient descent we converge to a very specific solution. Even though we were searching through an n-dimensional space, we converge to a unique point in this space.

This special w is the *minimum Euclidean norm solution* of $Xw = y$. In other words, out of all the linear prediction functions that interpolate the training data, SGD selects the solution with the minimal Euclidean norm. To see why this solution has minimal norm, suppose that $\widehat{w} = X^T\alpha + v$ with v orthogonal to all x_i. Then we have

$$X\widehat{w} = XX^T\alpha + Xv = XX^T\alpha\,,$$

which means α is completely determined and hence $\widehat{w} = X^T(XX^T)^{-1}y + v$. But now

$$\|\widehat{w}\|^2 = \|X^T(XX^T)^{-1}y\|^2 + \|v\|^2\,.$$

Minimizing the right-hand side shows that v must equal zero.

We now turn to showing that such minimum norm solutions have important robustness properties that suggest that they will perform well on new data. In the next chapter, we will prove that these methods are guaranteed to perform well on new data under reasonable assumptions.

Margin and stability

Consider a linear predictor that makes no classification errors and hence perfectly separates the data. Recall that the *decision boundary* of this predictor is the hyperplane $\mathcal{B} = \{z : w^T z = 0\}$ and the *margin* of the predictor is the distance of the decision boundary from the data:

$$\operatorname{margin}(w) = \min_i \operatorname{dist}(x_i, \mathcal{B})\,.$$

Since we're assuming that w correctly classifies all of the training data, we can write the margin in the convenient form

$$\operatorname{margin}(w) = \min_i \frac{y_i w^T x_i}{\|w\|}\,.$$

Ideally, we'd like our data to be far away from the boundary and hence we would like our predictor to have large margin. The reasoning behind this desideratum is as follows: If we expect new data to be similar to the training data and the decision boundary is far away from the training data, then it would be unlikely for a new data point to lie on the wrong side of the decision boundary. Note that the margin tells us how large a perturbation in the x_i can be handled before a data point is misclassified. It is a robustness measure that tells us how sensitive a predictor is to changes in the data itself.

Let's now specialize margin to the interpolation regime described in the previous section. Under the assumption that we interpolate the labels so that $w^T x_i = y_i$, we have

$$\operatorname{margin}(w) = \|w\|^{-1}\,.$$

If we want to simultaneously maximize margin and interpolate the data, then the optimal solution is to choose the minimum norm solution of $Xw = y$. This is precisely the solution found by SGD and gradient descent.

Note that we could have directly tried to maximize margin by solving the constrained optimization problem

$$\begin{array}{ll} \text{minimize} & \|w\|^2 \\ \text{subject to} & y_i w^T x_i \geq 1\,. \end{array}$$

This optimization problem is the classic formulation of the support vector machine. The support vector machine is an example of *explicit* regularization. Here we declare exactly which solution we'd like to choose given that our training error is zero. Explicit regularization of high-dimensional models is as old as machine learning. In contemporary machine learning, however, we often have to squint to see how our algorithmic decisions are regularizing. The trade-off is that we can run faster algorithms with implicit regularizers. But it's likely that revisiting classic regularization ideas in the context of contemporary models will lead to many new insights.

The representer theorem and kernel methods

As we have discussed so far, it is common in linear methods to restrict the search space to the span of the data. Even when d is large (or even infinite), this reduces the search problem to one in an n-dimensional space. It turns out that under broad generality, solutions in the span of the data are optimal for most optimization problems in prediction. Here, we make formal an argument we first introduced in our discussion of features: for most empirical risk minimization problems, the optimal model will lie in the span of the training data.

Consider the *penalized* ERM problem

$$\text{minimize} \frac{1}{n} \sum_{i=1}^{n} loss(w^T x_i, y_i) + \lambda \|w\|^2\,.$$

Here λ is called a *regularization parameter*. When $\lambda = 0$, there are an infinite number of w that minimize the ERM problem. But for any $\lambda > 0$, there is a unique minimizer. The term regularization refers to adding some prior information to an optimization problem to make the optimal solution unique. In this case, the prior information is explicitly encoding that we should prefer w with smaller norms if possible. As we discussed in our chapter on features, smaller norms tend to correspond to simpler solutions in many feature spaces. Moreover, we just described that minimum norm solutions themselves could be of interest in machine learning. A regularization parameter allows us to explicitly tune the norm of the optimal solution.

For our penalized ERM problem, using the same argument as above, we can write any w as

$$w = X^T\beta + v$$

for some vectors β and v with $v^T x_i = 0$ for all i. Plugging this equation into the penalized ERM problem yields

$$\text{minimize}_{\beta,v} \frac{1}{n}\sum_{i=1}^{n} loss(\beta^T X x_i, y_i) + \lambda\|X^T\beta\|^2 + \lambda\|v\|^2 .$$

Now we can minimize with respect to v, seeing that the only option is to set $v = 0$. Hence, we must have that the optimum model lies in the span of the data:

$$w = X^T\beta .$$

This derivation is commonly called the *representer theorem* in machine learning. As long as the cost function only depends on function evaluations $f(x_i) = w^T x_i$ and the cost increases as a function of $\|w\|$, then the empirical risk minimizer will lie in the span of the data.

Define the kernel matrix of the training data $K = XX^T$. We can then rewrite the penalized ERM problem as

$$\text{minimize}_{\beta} \frac{1}{n}\sum_{i=1}^{n} loss(e_i^T K\beta, y_i) + \lambda\beta^T K\beta ,$$

where e_i is the standard Euclidean basis vector. Hence, we can solve the machine learning problem only using the values in the matrix K, searching only for the coefficients β in the kernel expansion.

The representer theorem (also known as the kernel trick) tells us that most machine learning problems reduce to a search in n-dimensional space, even if the feature space has much higher dimension. Moreover, the optimization problems only care about the values of dot products between data points. This motivates the use of the kernel functions described in our discussion of representation: kernel functions allow us to evaluate dot products of vectors in high dimensional spaces often without ever materializing the vectors, reducing high-dimensional function spaces to the estimation of weightings of individual data points in the training sample.

Squared loss methods and other optimization tools

This chapter focused on gradient methods for minimizing empirical risk, as these are the most common methods in contemporary machine learning. However, there are a variety of other optimization methods that may be useful depending on the computational resources available and the particular application in question.

There are a variety of optimization methods that have proven fruitful in machine learning, most notably constrained quadratic programming for solving support vector machines and related problems. In this section we highlight least squares methods, which are attractive as they can be solved by solving linear systems. For many problems, linear systems solvers are faster than iterative gradient methods, and the computed solution is exact up to numerical precision, rather than being approximate.

Consider the optimization problem

$$\text{minimize}_w \quad \tfrac{1}{2}\textstyle\sum_{i=1}^n (y_i - w^T x_i)^2 .$$

The gradient of this loss function with respect to w is given by

$$-\sum_{i=1}^n (y_i - w^T x_i) x_i .$$

If we let y denote the vector of y labels and X denote the $n \times d$ matrix

$$X = \begin{bmatrix} x_1^T \\ x_2^T \\ \vdots \\ x_n^T \end{bmatrix},$$

then setting the gradient of the least squares cost equal to zero yields the solution

$$w = (X^T X)^{-1} X^T y .$$

For many problems, it is faster to compute this closed form solution than it is to run the number of iterations of gradient descent required to find a w with small empirical risk.

Regularized least squares also has a convenient closed form solution. The penalized ERM problem where we use a square loss is called the *ridge regression problem*:

$$\text{minimize}_w \quad \tfrac{1}{2}\textstyle\sum_{i=1}^n (y_i - w^T x_i)^2 + \lambda \|w\|^2 .$$

Ridge regression can be solved in the same manner as above, yielding the optimal solution

$$w = (X^T X + \lambda I)^{-1} X^T y .$$

There are a few other important problems solvable by least squares. First, we have the identity

$$(X^T X + \lambda I_d)^{-1} X^T = X(XX^T + \lambda I_n)^{-1} .$$

This means that we can solve ridge regression either by solving a system in d equations and d unknowns or n equations and n unknowns. In the overparameterized regime, we'd choose the formulation with n parameters. Moreover, as we described above, the minimum norm interpolating problem

$$\begin{array}{ll} \text{minimize} & \|w\|^2 \\ \text{subject to} & w^T x_i = y_i \end{array}$$

is solved by $w = X(XX^T)^{-1}y$. This shows that the limit as λ goes to zero in ridge regression is this minimum norm solution.

Finally, we note that for kernelized problems, we can simply replace the matrix XX^T with the appropriate kernel K. Hence, least squares formulations are extensible to solve prediction problems in arbitrary kernel spaces.

Chapter notes

Mathematical optimization is a vast field, and we clearly are only addressing a very small piece of the puzzle. For an expanded coverage of the material in this chapter with more mathematical rigor and implementation ideas, we invite the reader to consult the recent book by Wright and Recht.[80]

The chapter focuses mostly on iterative, stochastic gradient methods. Initially invented by Robbins and Monro for solving systems of equations in random variables,[76] stochastic gradient methods have played a key role in pattern recognition since the perceptron. Indeed, it was very shortly after Rosenblatt's invention that researchers realized the perceptron was solving a stochastic approximation problem. Of course, the standard perceptron step size schedule does not converge to a global optimum when the data is not separable, and this led to a variety of methods to fix the problem. Many researchers employed the Widrow-Hoff "Least-Mean-Squares" rule which in modern terms is minimizing the empirical risk associated with a square loss by stochastic gradient descent.[81] Aizerman and his colleagues determined not only how to apply stochastic gradient descent to linear functions, but how to operate in kernel spaces as well.[46] Of course, all of these methods were closely related to each other, but it took some time to put them all on a unified footing. It wasn't until the 1980s thath, with a full understanding of complexity theory, optimal step sizes were discovered for stochastic gradient methods by Nemirovski and Yudin.[82] More surprisingly, it was not until 2007 that the first non-asymptotic analysis of the perceptron algorithm was published.[83]

Interestingly, it wasn't again until the early 2000s that stochastic gradient descent became the default optimization method for machine learning. There tended to be a repeated cycle of popularity for the various optimization methods. Global optimization methods like linear programming were determined effective in the 1960s for pattern recognition problems,[45] supplanting interest in stochastic descent methods. Stochastic gradient descent was rebranded as back propagation in the 1980s, but again more global methods eventually took center stage. Mangasarian, who was involved in both of these cycles, told us in private correspondence that linear programming methods were always more effective in terms of their speed of computation and quality of solution.

Indeed this pattern was also followed in optimization. Nemirovski and Nesterov did pioneering work in iterative gradient and stochastic gradient methods . But they soon turned to developing the foundations of interior point methods for solving global optimization problems.[84] In the 2000s, they republished their work on iterative methods, leading to a revolution in machine learning.

It's interesting to track this history and forgetting in machine learning. Though these tools are not new, they are often forgotten and replaced. It's perhaps time to revisit the non-iterative methods in light of this.

Chapter 6

Generalization

Simply put, generalization relates the performance of a model on seen examples to its performance on *unseen* examples. In this chapter, we discuss the interplay between representation, optimization, and generalization, again focusing on models with more parameters than seen data points. We examine the intriguing empirical phenomena related to overparameterization and generalization in today's machine learning practice. We then review available theory—some old and some emerging—to better understand and anticipate what drives generalization performance.

Generalization gap

Recall, the risk of a predictor $f\colon \mathcal{X} \to \mathcal{Y}$ with respect to a loss function $loss\colon \mathcal{Y} \times \mathcal{Y} \to \mathbb{R}$ is defined as

$$R[f] = \mathbb{E}\left[loss(f(X), Y)\right] .$$

Throughout this chapter, it will often be convenient to stretch the notation slightly by using $loss(f, (x, y))$ to denote the loss of a predictor f on an example (x, y). For predictors specified by model parameters w, we'll also write $loss(w, (x, y))$.

For the purposes of this chapter, it makes sense to think of the n samples as an ordered tuple

$$S = ((x_1, y_1), \ldots\ldots, (x_n, y_n)) \in (\mathcal{X} \times \mathcal{Y})^n .$$

The empirical risk $R_S[f]$ is, as before,

$$R_S[f] = \frac{1}{n} \sum_{i=1}^{n} loss(f(x_i), y_i) .$$

Empirical risk minimization seeks to find a predictor f^* in a specified class $\mathcal{F}$ that minimizes the empirical risk:

$$R_S[f^*] = \min_{f \in \mathcal{F}} R_S[f].$$

In machine learning practice, the empirical risk is often called *training error* or *training loss*, as it corresponds to the loss achieved by some optimization method on the sample. Depending on the optimization problem, we may not be able to find an exact empirical risk minimizer and it may not be unique.

Empirical risk minimization is commonly used as a proxy for minimizing the unknown population risk. But how good is this proxy? Ideally, we would like that the predictor f we find via empirical risk minimization satisfy $R_S[f] \approx R[f]$. However, this may not be the case, since the risk $R[f]$ captures loss on unseen examples, while the empirical risk $R_S[f]$ captures loss on seen examples.

Generally, we expect to do much better on seen examples than unseen examples. This performance gap between seen and unseen examples is what we call *generalization gap*.

Definition 8. *Define the* generalization gap *of a predictor f with respect to a dataset S as*

$$\Delta_{\text{gen}}(f) = R[f] - R_S[f].$$

This quantity is sometimes also called *generalization error* or *excess risk*. Recall the following tautological, yet important identity:

$$R[f] = R_S[f] + \Delta_{\text{gen}}(f).$$

What it says is that if we manage to make the empirical risk $R_S[f]$ small through optimization, then all that remains to worry about is the generalization gap.

The last chapter provided powerful tools to make optimization succeed. How we can bound the generalization gap is the topic of this chapter. We first take a tour of evidence from machine learning practice for inspiration.

Overparameterization: Empirical phenomena

We previously experienced the advantages of overparameterized models in terms of their ability to represent complex functions and our ability to feasibly optimize them. The question remains whether they generalize well to unseen data. Perhaps we simply kicked the can down the road. Does the model size that was previously a blessing now come back to haunt us? We will see that not only do large models often generalize well in practice, but often more parameters lead to better generalization performance. Model size does, however, challenge some theoretical analysis. The empirical evidence will orient our theoretical study toward dimension-free bounds that avoid worst-case analysis.

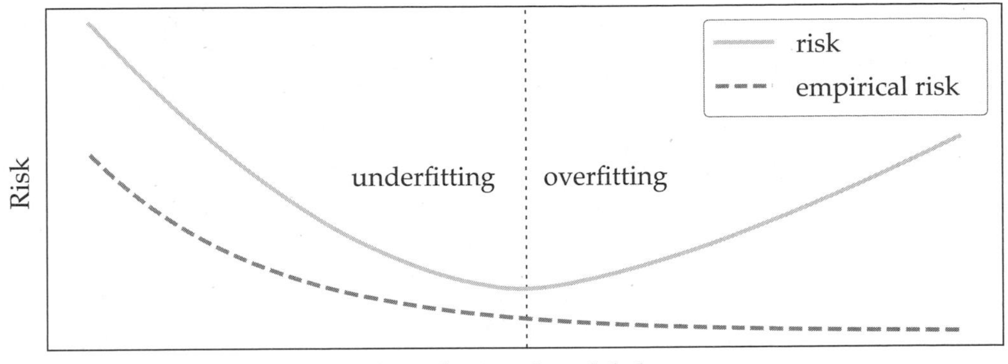

Figure 6.1: Traditional view of generalization

Effects of model complexity

Think of a model family with an associated measure of complexity, such as number of trainable parameters. Suppose that for each setting of the complexity measure, we can solve the empirical risk minimization problem. We can then plot what happens to risk and empirical risk as we vary model complexity.

A traditional view of generalization posits that as we increase model complexity, initially both empirical risk and risk decrease. However, past a certain point the risk begins to increase again, while empirical risk decreases.

The graphic shown in many textbooks is a U-shaped risk curve. The complexity range below the minimum of the curve is called *underfitting*. The range above is called *overfitting*.

This picture is often justified using the bias-variance trade-off, motivated by a least squares regression analysis. However, it does not seem to bear much resemblance to what is observed in practice.

We have already discussed the example of the perceptron, which achieves zero training loss and still generalizes well in theory. Numerous practitioners have observed that other complex models also can simultaneously achieve close to zero training loss and still generalize well. Moreover, in many cases risk continues to decreases as model complexity grows and training data are interpolated exactly down to (nearly) zero training loss. This empirical relationship between overparameterization and risk appears to be robust and manifests in numerous model classes, including overparameterized linear models, ensemble methods, and neural networks.

In the absence of regularization and for certain model families, the empirical relationship between model complexity and risk is more accurately captured by the *double descent* curve in the figure above. There is an interpolation threshold at which a model of the given complexity can fit the training data exactly. The complexity range below the threshold is the *underparameterized regime*, while

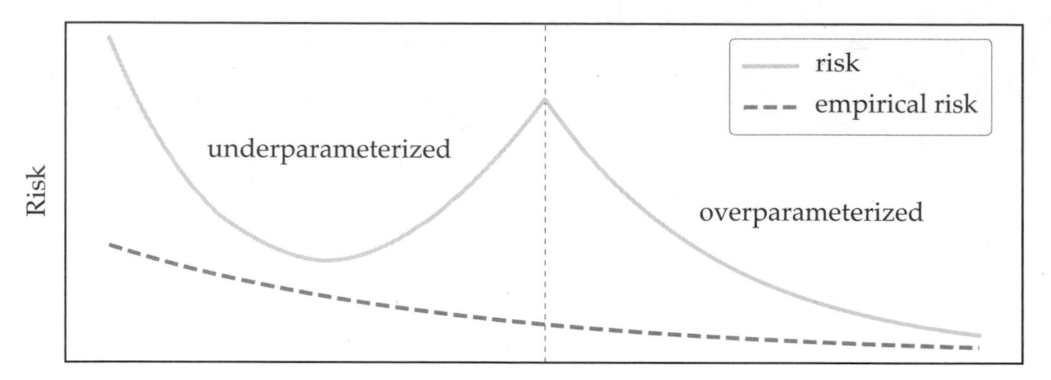

Figure 6.2: Double descent

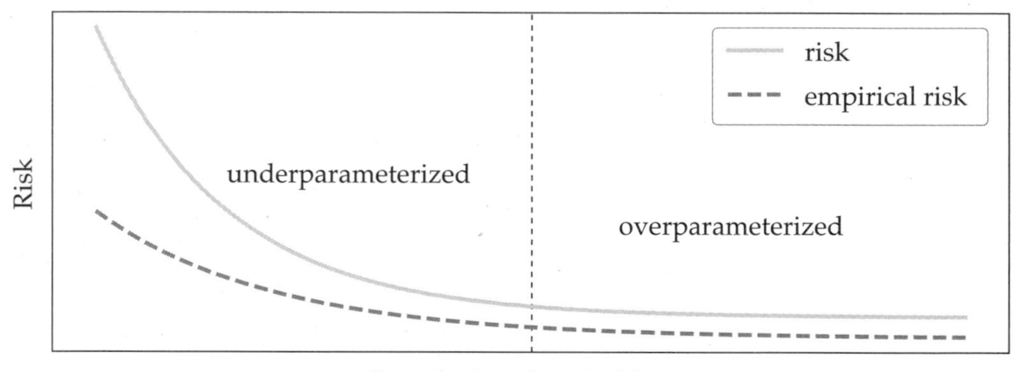

Figure 6.3: Single descent

the one above is the overparameterized regime. Increasing model complexity in the overparameterized regime continues to decrease risk indefinitely, albeit at decreasing marginal returns, toward some convergence point.

The double descent curve is not universal. In many cases, in practice we observe a single descent curve throughout the entire complexity range. In other cases, we can see multiple bumps as we increase model complexity.[85] However, the general point remains. There is no evidence that highly overparameterized models do not generalize. Indeed, empirical evidence suggests larger models not only generalize, but that larger models make better out-of-sample predictors than smaller ones.[86,87]

Optimization versus generalization

Training neural networks with stochastic gradient descent, as is commonly done in practice, attempts to solve a non-convex optimization problem. Reasoning

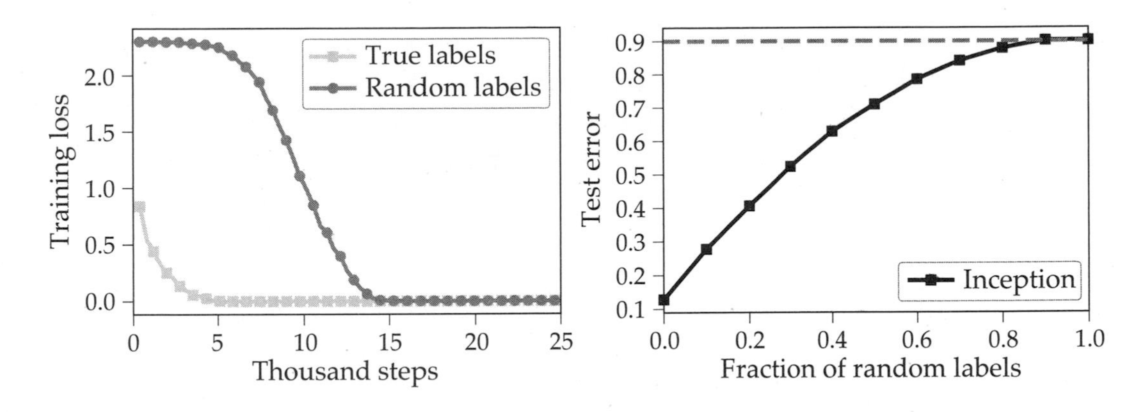

Figure 6.4: Randomization test on CIFAR-10. Left: How randomization affects training loss. Right: How increasing the fraction of corrupted training labels affects test error.

about non-convex optimization is known to be difficult. Theoreticians see a worthy goal in trying to prove mathematically that stochastic gradient methods successfully minimize the training objective of large artificial neural networks. The previous chapter discussed some of the progress that has been made toward this goal.

It is widely believed that what makes optimization easy crucially depends on the fact that models in practice have many more parameters than there are training points. While making optimization tractable, overparameterization puts burden on generalization.

We can force a disconnect between optimization and generalization in a simple experiment that we will see next. One consequence is that even if a mathematical proof established the convergence guarantees of stochastic gradient descent for training some class of large neural networks, it would not necessarily on its own tell us much about why the resulting model generalizes well to the test objective.

Indeed, consider the following experiment. Fix training data $(x_1, y_1), \ldots, (x_n, y_n)$ and fix a training algorithm A that achieves zero training loss on these data and achieves good test loss as well.

Now replace all the labels $y_1, \ldots, y_n$ by randomly and independently drawn labels $\tilde{y}_1, \ldots, \tilde{y}_n$. What happens if we run the same algorithm on the training data with noisy labels $(x_1, \tilde{y}_1), \ldots, (x_n, \tilde{y}_n))$?

One thing is clear. If we choose from k discrete classes, we expect the model trained on the random labels to have no more than $1/k$ test accuracy, that is, the accuracy achieved by random guessing. After all, there is no statistical relationship between the training labels and the test labels that the model could learn.

What is more interesting is what happens to optimization. The left panel of the figure shows the outcome of this kind of *randomization test* on the popular CIFAR-10 image classification benchmark for a standard neural network architecture. What we can see is that the training algorithm continues to drive the training loss to zero even if the labels are randomized. The right panel shows that we can vary the amount of randomization to obtain a smooth degradation of the test error. At full randomization, the test error degrades to 90%, as good as guessing one of the 10 classes. The figure shows what happens to a specific model architecture, called Inception, but similar observations hold for most, if not all, overparameterized architectures that have been proposed.

The randomization experiment shows that optimization continues to work well even when generalization performance is no better than random guessing, i.e., 10% accuracy in the case of the CIFAR-10 benchmark, which has 10 classes. The optimization method is moreover insensitive to properties of the data, since it works even on random labels. A consequence of this simple experiment is that a proof of convergence for the optimization method may not reveal any insights into the nature of generalization.

The diminished role of explicit regularization

Regularization plays an important role in the theory of convex empirical risk minimization. The most common form of regularization used to be ℓ_2-regularization, corresponding to adding a scalar of the squared Euclidean norm of the parameter vector to the objective function.

A more radical form of regularization, called *data augmentation*, is common in the practice of deep learning. Data augmentation transforms each training point repeatedly throughout the training process by some operation, such as a *random crop* of the image. Training on such randomly modified data points is meant to reduce overfitting, since the model never encounters the exact same data point twice.

Regularization continues to be a component of training large neural networks in practice. However, the nature of regularization is not clear. We can see a representative empirical observation in the table below.

Table 6.1: The training and test accuracy (in percentage) with and without data augmentation and ℓ_2-regularization.

params	random crop	ℓ_2-regularization	train accuracy	test accuracy
1,649,402	yes	yes	100.0	89.05
	yes	no	100.0	89.31
	no	yes	100.0	86.03
	no	no	100.0	85.75

The table shows the performance of a common neural model architecture, called Inception, on the standard CIFAR-10 image classification benchmark. The model has more than 1.5 million trainable parameters, even though there are only $50,000$ training examples spread across 10 classes. The training procedure uses two explicit forms of regularization. One is a form of data augmentation with random crops. The other is ℓ_2-regularization. With both forms of regularization the fully trained model achieves close to 90% test accuracy. But even if we turn both of them off, the model still achieves close to 86% test accuracy (without even readjusting any hyperparameters such as learning rate of the optimizer). At the same time, the model fully interpolates the training data in the sense of making no errors whatsoever on the training data.

These findings suggest that while explicit regularization may help generalization performance, it is by no means necessary for strong generalization of heavily overparameterized models.

Theories of generalization

With these empirical facts in hand, we now turn to mathematical theories that might help explain what we observe in practice and also may guide future empirical and theoretical work. In the remainder of the chapter, we tour several different, seemingly disconnected views of generalization.

We begin with a deep dive into *algorithmic stability*, which posits that generalization arises when models are insensitive to perturbations in the data on which they are trained. We then discuss *VC dimension and Rademacher complexity*, which show how small generalization gaps can arise when we restrict the complexity of models we wish to fit to data. We then turn to *margin bounds*, which assert that whenever the data is easily separable, good generalization will occur. Finally we discuss generalization bounds that arise from *optimization*, showing how choice of an algorithmic scheme itself can yield models with desired generalization properties.

In all of these cases, we show that we can recover generalization bounds of the form we saw in the perceptron: the bounds will decrease with number of data points and increase with "complexity" of the optimal prediction function. Indeed, looking back at the proof of the perceptron generalization bound, all of the above elements appeared. Our generalization bound arose because we could remove single data points from a set and not change the number of mistakes made by the Perceptron. A large margin assumption was essential to getting a small mistake bound. The mistake bound itself was dependent on the iterations of the algorithm. And finally, we related the size of the margin to the scale of the data and the optimal separator.

Though starting from different places, we will show that the four different views of generalization can all arrive at similar results. Each of the aforementioned ingredients can alone lead to generalization, but considerations of all of these aspects help to improve machine learning methods. Generalization is multifaceted and multiple perspectives are useful when designing data-driven predictive systems.

Before diving into these four different views, we first take a quick pause to consider how we hope generalization bounds might look.

How should we expect the gap to scale?

Before we turn to analyzing generalization gaps, it's worth first considering how we should expect them to scale. That is, what is the relationship between the expected size of Δ_{gen} and the number of observations, n?

First, note that for a *fixed* prediction function f, the expectation of the empirical risk is equal to the population risk. That is, the empirical risk of a single function is a sample average of the population risk of that function. As we discussed in Chapter 3, i.i.d. sample averages should *generalize* and approximate the mean at the population level. Here, we now turn to describing *how* they might be expected to scale under different assumptions.

Quantitative central limit theorems

The *central limit theorem* formalizes how sample averages estimate their expectations: If Z is a random variable with bounded variance, then $\widehat{\mu}_Z^{(n)}$ converges in distribution to a Gaussian random variable with mean zero and variance on the order of $1/n$.

The following inequalities are useful *quantitative* forms of the central limit theorem. They precisely measure how close the sample average will be to the population average using limited information about the random quantity.

- **Markov's inequality:** Let Z be a nonnegative random variable. Then for any $t > 0$,

$$\mathbb{P}[Z \geq t] \leq \frac{\mathbb{E}[Z]}{t}.$$

This can be proven using the inequality $\mathbb{1}\{z \geq t\} \leq \frac{z}{t}$, for $z \geq 0$.

- **Chebyshev's inequality:** Suppose Z is a random variable with mean μ and variance σ^2. Then for any $t > 0$,

$$\mathbb{P}[Z \geq t + \mu] \leq \frac{\sigma^2}{t^2}.$$

Chebyshev's inequality helps us understand why sample averages are good estimates of the mean. Suppose that $Z_1, \ldots, Z_n$ are independent random variables, each with mean μ and variance σ^2. Let $Z = \frac{1}{n}\sum_{i=1}^n Z_i$ denote the sample average. Chebyshev's inequality implies

$$\mathbb{P}[Z \geq t + \mu] \leq \frac{\sigma^2}{nt^2},$$

which tends to zero as n grows. A popular form of this inequality sets $t = \sigma$ which gives

$$\mathbb{P}[Z \geq \mu + \sigma] \leq \frac{1}{n}.$$

- **Hoeffding's inequality:** Let $Z_1, Z_2, \ldots, Z_n$ be independent random variables, each Z_i taking values in the interval $[a_i, b_i]$. Let $Z = \frac{1}{n}\sum_{i=1}^n Z_i$ denote the sample average. Then

$$\mathbb{P}[Z \geq \mathbb{E}[Z] + t] \leq \exp\left(-\frac{2n^2t^2}{\sum_{i=1}^n (b_i - a_i)^2}\right).$$

An important special case is when the Z_i are identically distributed and take values in $[0, 1]$. Then we have

$$\mathbb{P}[Z \geq \mathbb{E}[Z] + t] \leq \exp\left(-2nt^2\right).$$

This shows that when random variables are bounded, sample averages concentrate around their mean value exponentially quickly. If we invoke this bound with $t = C/\sqrt{n}$, the point at which it gives nontrivial results, we have an error of $O(1/\sqrt{n})$ with all but exponentially small probability. We will see shortly that this relationship between error and number of samples is ubiquitous in generalization theory.

These powerful *concentration inequalities* let us precisely quantify how close the sample average will be to the population mean. For instance, we know a person's height is a positive number and that there are no people who are taller than nine feet. With these two facts, Hoeffding's inequality tells us that if we sample the heights of 30,000 individuals, our sample average will be within an inch of the true average height with probability at least 83%. This assertion is true no matter how large the population of individuals. The required sample size is dictated only by the variability of height, not by the number of total individuals.

You could replace "height" in this example with almost any attribute that you are able to measure well. The quantitative central limits tell us that for attributes with reasonable variability, a uniform sample from a general population will give a high-quality estimate of the average value.

"Reasonable variability" of a random variable is necessary for quantitative central limit theorems to hold. When random variables have low variance or are tightly bounded, small experiments quickly reveal insights about the population. When variances are large or effectively unbounded, the number of samples required for high-precision estimates might be impractical and our estimators and algorithms and predictions may need to be rethought.

Bounding generalization gaps for individual predictors

Let us now return to generalization of prediction, considering the example where the quantity of interest is the prediction error on individuals in a population. There are effectively two scaling regimes of interest in generalization theory. In one case, when the empirical risk is large, we expect the generalization gap to decrease inversely proportional to $\sqrt{n}$. When the empirical risk is expected to be very small, on the other hand, we tend to see the the generalization gap decrease inversely proportional to n.

Why we see these two regimes is illustrated by studying the case of a *single* prediction function f, chosen *independently* of the sample S. Our ultimate goal is to reason about the generalization gap of predictors chosen by an algorithm running on our data. The analysis we walk through next doesn't apply to data-dependent predictors directly, but it nonetheless provides helpful intuition about what bounds we can hope to get.

For a fixed function f, the zero-one loss on a single randomly chosen data point is a Bernoulli random variable, equal to 1 with probability p and $1-p$ otherwise. The empirical risk $R_S[f]$ is the sample average of this random variable and the risk $R[f]$ is its expectation. To estimate the generalization gap, we can apply Hoeffding's inequality to find

$$\mathbb{P}[R[f] - R_S[f] \geq \epsilon] \leq \exp\left(-2n\epsilon^2\right) .$$

Hence, we will have with probability $1-\delta$ on our sample that

$$|\Delta_{\text{gen}}(f)| \leq \sqrt{\frac{\log(1/\delta)}{2n}} .$$

That is, the generalization gap goes to zero at a rate of $1/\sqrt{n}$.

In the regime where we observe no empirical mistakes, a more refined analysis can be applied. Suppose that $R[f] > \epsilon$. Then the probability that we observe $R_S[f] = 0$ cannot exceed

$$\begin{aligned}\mathbb{P}[\forall i\colon \operatorname{sign}(f(x_i)) = y_i] &= \prod_{i=1}^{n} \mathbb{P}[\operatorname{sign}(f(x_i)) = y_i] \\ &\leq (1-\epsilon)^n \leq e^{-\epsilon n} .\end{aligned}$$

Hence, with probability $1-\delta$,

$$|\Delta_{\text{gen}}(f)| \leq \frac{\log(1/\delta)}{n} ,$$

which is the $1/n$ regime. These two rates are precisely what we observe in the more complex regime of generalization bounds in machine learning. The main trouble and difficulty in computing bounds on the generalization gap is that our prediction function f depends on the data, making the above analysis inapplicable.

In this chapter, we will focus mostly on $1/\sqrt{n}$ rates. These rates are more general as they make no assumptions about the expected empirical risk. With a few notable exceptions, the derivation of $1/\sqrt{n}$ rates tends to be easier than that of the $1/n$ counterparts. However, we note that every one of our approaches to generalization bounds has analyses for the "low empirical risk" or "large margin" regimes. We provide references at the end of this chapter to these more refined analyses.

Algorithmic stability

We will first see a tight characterization in terms of an algorithmic robustness property we call *algorithmic stability*. Intuitively, algorithmic stability measures how sensitive an algorithm is to changes in a single training example. Whenever a model is insensitive to such perturbations, the generalization gap will be small. Stability gives us a powerful and intuitive way of reasoning about generalization.

There are a variety of different notions of perturbation. We could consider resampling a single data point and look at how much a model changes. We could also leave one data point out and see how much the model changes. This was the heart of our perceptron generalization argument. More aggressively, we could study what happens when a single data point is arbitrarily corrupted. All three of these approaches yield similar generalization bounds, though it is often easier to work with one than the others. To simplify the exposition, we choose to focus on only one notion (resampling) here.

To introduce the idea of stability, we first condense our notation to make the presentation a bit less cumbersome. Recall that we operate on tuples of n labeled examples,

$$S = ((x_1, y_1), \ldots\ldots, (x_n, y_n)) \in (\mathcal{X} \times \mathcal{Y})^n .$$

We denote a labeled example as $z = (x, y)$. We will overload our notation and denote the loss accrued by a prediction function f on a data point z as $loss(f, z)$. That is, $loss(f, z) = loss(f(x), y)$. We use the uppercase letters when a labeled example Z is randomly drawn from a population (X, Y).

With this notation in hand, let's now consider two independent random samples $S = (Z_1, \ldots, Z_n)$ and $S' = (Z_1', \ldots, Z_n')$, each drawn independently and identically from a population (X, Y). We call the second sample S' a *ghost sample* as it is solely an analytical device. We never actually collect this second sample or run any algorithm on it.

We introduce *n hybrid samples* $S^{(i)}$, for $i \in \{1, \ldots, n\}$ as

$$S^{(i)} = (Z_1, \ldots, Z_{i-1}, Z_i', Z_{i+1}, \ldots, Z_n),$$

where the ith example comes from S', while all others come from S.

We can now introduce a data-dependent notion of average stability of an algorithm. For this definition, we think of an algorithm as a deterministic map A that takes a training sample in $(\mathcal{X} \times \mathcal{Y})^n$ to some prediction function in a function space Ω. That is, $A(S)$ denotes the function from $\mathcal{X}$ to $\mathcal{Y}$ that is returned by our algorithm when run on the sample S.

Definition 9. *The* average stability *of an algorithm* $A\colon (\mathcal{X} \times \mathcal{Y})^n \to \Omega$ *is*

$$\Delta(A) = \mathop{\mathbb{E}}_{S,S'} \left[\frac{1}{n} \sum_{i=1}^{n} \left(loss(A(S), Z_i') - loss(A(S^{(i)}), Z_i') \right) \right].$$

There are two useful ways to parse this definition. The first is to interpret average stability as the average sensitivity of the algorithm to a change in a single example. Since we don't know which of its n input samples the algorithm may be sensitive to, we test all of them and average out the results.

Second, from the perspective of $A(S)$, the example Z_i' is *unseen*, since it is not part of S. But from the perspective of $A(S^{(i)})$ the example Z_i' is seen, since it is part of $S^{(i)}$ via the substitution that defines the ith hybrid sample. This shows that the instrument $\Delta(A)$ also measures the average loss difference of the algorithm on seen and unseen examples. We therefore have reason to suspect that average stability relates to generalization gap, as the next proposition confirms.

Proposition 7. *The expected generalization gap equals average stability:*

$$\mathbb{E}[\Delta_{\text{gen}}(A(S))] = \Delta(A).$$

Proof. By linearity of expectation,

$$\begin{aligned} \mathbb{E}[\Delta_{\text{gen}}(A(S))] &= \mathbb{E}\left[R[A(S)] - R_S[A(S)]\right] \\ &= \mathbb{E}\left[\frac{1}{n}\sum_{i=1}^{n} loss(A(S), Z_i')\right] - \mathbb{E}\left[\frac{1}{n}\sum_{i=1}^{n} loss(A(S), Z_i)\right]. \end{aligned}$$

Here, we used that Z_i' is an example drawn from the distribution that does not appear in the set S, while Z_i does appear in S. At the same time, Z_i and Z_i' are identically distributed and independent of the other examples. Therefore,

$$\mathbb{E}\, loss(A(S), Z_i) = \mathbb{E}\, loss(A(S^{(i)}), Z_i').$$

Applying this identity to each term in the empirical risk above, and comparing with the definition of $\Delta(A)$, we conclude

$$\mathbb{E}[R[A(S)] - R_S[A(S)]] = \Delta(A)\,. \qquad \square$$

Uniform stability

While average stability gave us an exact characterization of generalization error, it can be hard to work with the expectation over S and S'. Uniform stability replaces the averages by suprema, leading to a stronger but useful notion.

Definition 10. *The* uniform stability *of an algorithm A is defined as*

$$\Delta_{\sup}(A) = \sup_{\substack{S,S' \in (\mathcal{X}\times\mathcal{Y})^n \\ d_H(S,S')=1}} \sup_{z \in \mathcal{X}\times\mathcal{Y}} |loss(A(S),z) - loss(A(S'),z)|,$$

where $d_H(S,S')$ is the Hamming distance between tuples S and S'.

In this definition, it is important to note that the z has nothing to do with S and S'. Uniform stability is effectively computing the worst-case difference in the predictions of the learning algorithm run on two arbitrary datasets that differ in exactly one point.

Uniform stability upper bounds average stability, and hence uniform stability upper bounds generalization gap (in expectation). Thus, we have the corollary

$$\mathbb{E}[\Delta_{\text{gen}}(A(S))] \le \Delta_{\sup}(A)\,.$$

This corollary turns out to be surprisingly useful since many algorithms are uniformly stable. For example, strong convexity of the loss function is sufficient for the uniform stability of empirical risk minimization, as we will see next.

Stability of empirical risk minimization

We now show that empirical risk minimization is uniformly stable under strong assumptions on the loss function. One important assumption we need is that the loss function $loss(w,z)$ is differentiable and *strongly convex* in the model parameters w for every example z. What this means is that for every example z and for all $w, w' \in \Omega$,

$$loss(w',z) \ge loss(w,z) + \langle \nabla loss(w,z), w' - w\rangle + \frac{\mu}{2}\|w - w'\|^2\,.$$

There's only one property of strong convexity we'll need. Namely, if $\Phi\colon \mathbb{R}^d \to \mathbb{R}$ is μ-strongly convex and w^* is a stationary point (and hence global minimum) of the function Φ, then we have

$$\Phi(w) - \Phi(w^*) \ge \frac{\mu}{2}\|w - w^*\|^2\,.$$

The second assumption we need is that $loss(w,z)$ is *L-Lipschitz* in w for every z, i.e., $\|\nabla loss(w,z)\| \leq L$. Equivalently, this means $|loss(w,z) - loss(w',z)| \leq L\|w - w'\|$.

Theorem 4. *Assume that for every z, $loss(w,z)$ is μ-strongly convex in w over the domain Ω. Further, assume that the loss function $loss(w,z)$ is L-Lipschitz in w for every z. Then, empirical risk minimization (ERM) satisfies*

$$\Delta_{\text{sup}}(ERM) \leq \frac{4L^2}{\mu n}\,.$$

Proof. Let $\widehat{w}_S = \arg\min_{w\in\Omega} \frac{1}{n}\sum_{i=1}^{n} loss(w,z_i)$ denote the empirical risk minimizer on the sample S. Fix arbitrary samples S, S' of size n that differ in a single index $i \in \{1,\ldots,n\}$ where S contains z_i and S' contains z_i'. Fix an arbitrary example z. We need to show that

$$|loss(\widehat{w}_S, z) - loss(\widehat{w}_{S'}, z)| \leq \frac{4L^2}{\mu n}\,.$$

Since the loss function is L-Lipschitz, it suffices to show that

$$\|\widehat{w}_S - \widehat{w}_{S'}\| \leq \frac{4L}{\mu n}\,.$$

On the one hand, since $\widehat{w}_S$ minimizes the empirical risk by definition, it follows from the strong convexity of the empirical risk that

$$\frac{\mu}{2}\|\widehat{w}_S - \widehat{w}_{S'}\|^2 \leq R_S[\widehat{w}_{S'}] - R_S[\widehat{w}_S]\,.$$

On the other hand, we can bound the right-hand side as

$$\begin{aligned}
&R_S[\widehat{w}_{S'}] - R_S[\widehat{w}_S] \\
&= \frac{1}{n}(loss(\widehat{w}_{S'}, z_i) - loss(\widehat{w}_S, z_i)) + \frac{1}{n}\sum_{i\neq j}(loss(\widehat{w}_{S'}, z_j) - loss(\widehat{w}_S, z_j)) \\
&= \frac{1}{n}(loss(\widehat{w}_{S'}, z_i) - loss(\widehat{w}_S, z_i)) + \frac{1}{n}(loss(\widehat{w}_S, z_i') - loss(\widehat{w}_{S'}, z_i')) \\
&\quad + (R_{S'}[\widehat{w}_{S'}] - R_{S'}[\widehat{w}_S]) \\
&\leq \frac{1}{n}|loss(\widehat{w}_{S'}, z_i) - loss(\widehat{w}_S, z_i)| + \frac{1}{n}|loss(\widehat{w}_S, z_i') - loss(\widehat{w}_{S'}, z_i')| \\
&\leq \frac{2L}{n}\|\widehat{w}_{S'} - \widehat{w}_S\|\,.
\end{aligned}$$

Here, we used the assumption that $loss$ is L-Lipschitz and the fact that

$$R_{S'}[\widehat{w}_{S'}] - R_{S'}[\widehat{w}_S] \leq 0\,.$$

Putting together the strong convexity property and our calculation above, we find

$$\|\widehat{w}_{S'} - \widehat{w}_S\| \leq \frac{4L}{\mu n}.$$

Hence, $\Delta_{\sup}(\text{ERM}) \leq \frac{4L^2}{\mu n}$. □

An interesting point about this result is that there is no explicit reference to the complexity of the model class referenced by Ω.

Stability of regularized empirical risk minimization

Some empirical risk minimization problems, such as the perceptron (ERM with hinge loss) we saw earlier, are convex but not strictly convex. We can turn convex problems into strongly convex problems by adding an ℓ_2-*regularization* term to the loss function:

$$r(w,z) = loss(w,z) + \frac{\mu}{2}\|w\|^2.$$

The last term is named ℓ_2-regularization, *weight decay*, or *Tikhonov regularization* depending on field and context.

By construction, if the loss is convex, then the regularized loss $r(w,z)$ is μ-strongly convex. Hence, our previous theorem applies. However, by adding regularization we changed the objective function. The optimizer of the regularized objective is in general not the same as the optimizer of the unregularized objective.

Fortunately, a simple argument shows that solving the regularized objective also solves the unregularized objective. The idea is that assuming $\|w\| \leq B$ we can set the regularization parameter $\mu = \frac{L}{B\sqrt{n}}$. This ensures that the regularization term $\mu\|w\|^2$ is at most $O(\frac{LB}{\sqrt{n}})$ and therefore the minimizer of the regularized risk also minimizes the unregularized risk up to error $O(\frac{LB}{\sqrt{n}})$. Plugging this choice of μ into the ERM stability theorem, the generalization gap will also be $O(\frac{LB}{\sqrt{n}})$.

The case of regularized hinge loss

Let's relate the generalization theory we just saw to the familiar case of the perceptron algorithm from Chapter 3. This corresponds to the special case of minimizing the regularized hinge loss

$$r(w,(x,y)) = \max\{1 - y\langle w,x\rangle, 0\} + \frac{\mu}{2}\|w\|^2.$$

Moreover, we assume that the data are are linearly separable with margin γ.

Denoting by $\widehat{w}_S$ the empirical risk minimizer on a random sample S of size n, we know that

$$\frac{\mu}{2}\|\widehat{w}_S\|^2 \leq R_S(\widehat{w}_S) \leq R_S(0) = 1\,.$$

Hence, $\|\widehat{w}_S\| \leq B$ for $B = \sqrt{2/\mu}$. We can therefore restrict our domain to the Euclidean ball of radius B. If the data are also bounded, say $\|x\| \leq D$, we further get that

$$\|\nabla_w r(w,z)\| \leq \|x\| + \mu\|w\| = D + \mu B\,.$$

Hence, the regularized hinge loss is L-Lipschitz with

$$L = D + \mu B = D + \sqrt{2\mu}\,.$$

Let w_γ be a maximum margin hyperplane for the sample S. We know that the empirical loss will satisfy

$$R_S[\widehat{w}_S] \leq R_S[w_\gamma] = \frac{\mu}{2}\|w_\gamma\|^2 = \frac{\mu}{2\gamma^2}\,.$$

Hence, by Theorem 4,

$$\mathbb{E}[R[\widehat{w}_S]] \leq \mathbb{E}[R_S[\widehat{w}_S]] + \Delta_{\text{sup}}(\text{ERM}) \leq \frac{\mu}{2\gamma^2} + \frac{4(D+\sqrt{2\mu})^2}{\mu n}\,.$$

Setting $\mu = \frac{2\gamma D}{\sqrt{n}}$ and noting that $\gamma \leq D$ gives that

$$\mathbb{E}[R[\widehat{w}_S]] \leq O\left(\frac{D}{\gamma\sqrt{n}}\right)\,.$$

Finally, since the regularized hinge loss upper bounds the zero-one loss, we can conclude that

$$\mathbb{P}[Y\widehat{w}_S^T X < 0] \leq O\left(\frac{D}{\gamma\sqrt{n}}\right)\,,$$

where the probability is taken over both sample S and test point (X,Y). Applying Markov's inequality to the sample, we can conclude the same bound holds for a typical sample up to constant factors.

This bound is proportional to the square root of the bound we saw for the perceptron in Chapter 3. As we discussed earlier, this rate is slower than the perceptron rate as it does not explicitly take into account the fact that the empirical risk is zero. However, it is worth noting that the relationship between the variables in question—diameter, margin, and number of samples—is precisely the same as for the perceptron. This kind of bound is common and we will derive it a few more times in this chapter.

Stability analysis combined with explicit regularization and convexity thus give an appealing conceptual and mathematical approach to reasoning about generalization. However, empirical risk minimization involving non-linear models is increasingly successful in practice and generally leads to non-convex optimization problems.

Model complexity and uniform convergence

We briefly review other useful tools to reason about generalization. Arguably, the most basic is based on counting the number of different functions that can be described with the given model parameters.

Given a sample S of n independent draws from the same underlying distribution, the empirical risk $R_S[f]$ for a fixed function f is an average of n random variables, each with mean equal to the risk $R[f]$. Assuming for simplicity that the range of our loss function is bounded in the interval $[0,1]$, Hoeffding's bound gives us the tail bound

$$\mathbb{P}\left[R_S[f] > R[f] + t\right] \leq \exp(-2nt^2).$$

By applying the union bound to a finite set of functions $\mathcal{F}$ we can guarantee that with probability $1-\delta$ we have for all functions $f \in \mathcal{F}$ that

$$\Delta_{\text{gen}}(f) \leq \sqrt{\frac{\ln|\mathcal{F}| + \ln(1/\delta)}{n}}. \qquad (2)$$

The cardinality bound $|\mathcal{F}|$ is a basic measure of the complexity of the model family $\mathcal{F}$. We can think of the term $\ln(\mathcal{F})$ as a measure of complexity of the function class $\mathcal{F}$. The gestalt of the generalization bound as "$\sqrt{\text{complexity}/n}$" routinely appears with varying measures of complexity.

VC dimension

Bounding the generalization gap from above for all functions in a function class is called *uniform convergence*. A classical tool to reason about uniform convergence is the Vapnik-Chervonenkis dimension (VC dimension) of a function class $\mathcal{F} \subseteq X \to Y$, denoted $\text{VC}(\mathcal{F})$. It's defined as the size of the largest set $Q \subseteq X$ such that for any Boolean function $h\colon Q \to \{-1,1\}$, there is a predictor $f \in \mathcal{F}$ such that $f(x) = h(x)$ for all $x \in Q$. In other words, if there is a size-d sample Q such that the functions of $\mathcal{F}$ induce all 2^d possible binary labelings of Q, then the VC-dimension of $\mathcal{F}$ is at least d.

The VC-dimension measures the ability of the model class to conform to an arbitrary labeling of a set of points. The so-called VC inequality implies that with probability $1-\delta$, we have for all functions $f \in \mathcal{F}$

$$\Delta_{\text{gen}}(f) \leq \sqrt{\frac{\text{VC}(\mathcal{F})\ln n + \ln(1/\delta)}{n}}. \qquad (3)$$

We can see that the complexity term $\text{VC}(\mathcal{F})$ refines our earlier cardinality bound since $\text{VC}(\mathcal{F}) \leq \log|\mathcal{F}| + 1$. However VC-dimension also applies to infinite model classes. Linear models over $\mathbb{R}^d$ have VC-dimension d, corresponding to the number of model parameters. Generally speaking, VC dimension tends to grow with the number of model parameters for many model families of interest. In such cases, the bound in Equation 3 becomes useless once the number of model parameters exceeds the size of the sample.

However, the picture changes significantly if we consider notions of model complexity different than raw counts of parameters. Consider the set of all hyperplanes in $\mathbb{R}^d$ with norm at most γ^{-1}. Vapnik showed[88] that when the data had maximum norm $\|x\| \leq D$, then the VC dimension of this set of hyperplanes was $\frac{D^2}{\gamma^2}$. As described in a survey of support vector machines by Burges, the worst-case arrangement of n data points is a simplex in $n-2$ dimensions.[89] Plugging this VC-dimension into our generalization bound yields

$$\Delta_{\text{gen}}(f) \leq \sqrt{\frac{D^2 \ln n + \gamma^2 \ln(1/\delta)}{\gamma^2 n}}\,.$$

We again see our perceptron style generalization bound! This bound again holds when the empirical risk is nonzero. The dimension of the data, d, does not appear at all in this bound. The difference between the parametric model and the margin-like bound is that we considered properties of the data. In the *worst-case* bound that counts parameters, it appears that high-dimensional prediction is impossible. It is only by considering data-specific properties that we can find a reasonable generalization bound.

Rademacher complexity

An alternative to VC-dimension is Rademacher complexity, a flexible tool that often is more amenable to calculations that incorporate problem-specific aspects such as restrictions on the distribution family or properties of the loss function. To get a generalization bound in terms of Rademacher complexity, we typically apply the definition not to the model class $\mathcal{F}$ itself but to the class $\mathcal{L}$ of functions of the form $h(z) = \mathit{loss}(f, z)$ for some $f \in \mathcal{F}$ and a loss function loss. By varying the loss function, we can derive different generalization bounds.

Fix a function class $\mathcal{L} \subseteq Z \to \mathbb{R}$, which will later correspond to the composition of a predictor with a loss function, which is why we chose the symbol $\mathcal{L}$. Think of the domain Z as the space of labeled examples $z = (x, y)$. Fix a distribution P over the space Z.

The *empirical Rademacher complexity* of a function class $\mathcal{L} \subseteq Z \to \mathbb{R}$ with respect to a sample $\{z_1, \ldots, z_n\} \subseteq Z$ drawn i.i.d. from the distribution P is defined as:

$$\widehat{\mathfrak{R}}_n(\mathcal{L}) = \mathop{\mathbb{E}}_{\sigma \in \{-1,1\}^n} \left[\frac{1}{n} \sup_{h \in \mathcal{L}} \left| \sum_{i=1}^{n} \sigma_i h(z_i) \right| \right].$$

We obtain the *Rademacher complexity* $\mathfrak{R}_n(\mathcal{L}) = \mathbb{E}\left[\widehat{\mathfrak{R}}_n(\mathcal{L})\right]$ by taking the expectation of the empirical Rademacher complexity with respect to the sample.

Rademacher complexity measures the ability of a function class to interpolate a random sign pattern assigned to a point set.

One application of Rademacher complexity applies when the loss function is L-Lipschitz in the parameterization of the model class for every example z. This bound shows that with probability $1-\delta$ for all functions $f \in \mathcal{F}$, we have

$$\Delta_{\text{gen}}(f) \le 2L\mathfrak{R}_n(\mathcal{F}) + 3\sqrt{\frac{\log(1/\delta)}{n}}.$$

When applied to the hinge loss with the function class being hyperplanes of norm less than γ^{-1}, this bound again recovers the perceptron generalization bound:

$$\Delta_{\text{gen}}(f) \le \frac{2D}{\gamma\sqrt{n}} + 3\sqrt{\frac{\log(1/\delta)}{n}}.$$

Margin bounds for ensemble methods

Ensemble methods work by combining many weak predictors into one strong predictor. The combination step usually involves taking a weighted average or majority vote of the weak predictors. Boosting and random forests are two ensemble methods that continue to be highly popular and competitive in various settings. Both methods train a sequence of small decision trees, each on its own achieving modest accuracy on the training task. However, so long as different trees make errors that aren't too correlated, we can obtain a higher accuracy model by taking, say, a majority vote over the individual predictions of the trees.

Researchers in the 1990s observed that boosting often continues to improve test accuracy as more weak predictors are added to the ensemble. The complexity of the entire ensemble was thus often far too large to apply standard uniform convergence bounds.

A proffered explanation was that boosting, while growing the complexity of the ensemble, also improved the *margin* of the ensemble predictor. Assuming that the final predictor $f\colon X \to \{-1,1\}$ is binary, its *margin* on an example (x,y) is defined as the value $yf(x)$. The larger the margin the more "confident" the predictor is about its prediction. A margin $yf(x)$ just above 0 shows that the weak predictors in the ensemble were nearly split evenly in their weighted votes.

An elegant generalization bound relates the risk of any predictor f to the fraction of correctly labeled training examples at a given margin θ. Below, let $R[f]$ be the risk of f w.r.t. zero-one loss. However, let $R_S^\theta(f)$ be the empirical risk with respect to *margin errors* at level θ, i.e., the loss $\mathbb{1}\{yf(x) \le \theta\}$ that penalizes errors where the predictor is within an additive θ margin of making a mistake.

Theorem 5. *With probability $1-\delta$, every convex combination f of base predictors in $\mathcal{H}$ satisfies the following bound for every $\theta > 0$:*

$$R[f] - R_S^\theta[f] \le O\left(\frac{1}{\sqrt{n}}\left(\frac{\mathrm{VC}(\mathcal{H})\log n}{\theta^2} + \log(1/\delta)\right)^{1/2}\right).$$

The theorem can be proved using Rademacher complexity. Crucially, the bound depends only on the VC dimension of the base class $\mathcal{H}$, not the complexity of ensemble. Moreover, the bound holds for all $\theta > 0$, and so we can choose θ after knowing the margin that manifested during training.

Margin bounds for linear models

Margins also play a fundamental role for linear prediction. We saw one margin bound for linear models in our chapter on the perceptron algorithm. Similar bounds hold for other variants of linear prediction. We'll state the result here for a simple least squares problem:

$$w^* = \arg\min_{w:\ \|w\|\le B} \frac{1}{n}\sum_{i=1}^{n} (\langle x_i, w\rangle - y)^2 .$$

In other words, we minimize the empirical risk w.r.t. the squared loss over norm bounded linear separators; call this class $\mathcal{W}_B$. Further assume that all data points satisfy $\|x_i\| \le 1$ and $y \in \{-1, 1\}$. Analogous to the margin bound in Theorem 5, it can be shown that with probability $1-\delta$ for every linear predictor f specified by weights in $\mathcal{W}_B$ we have

$$R[f] - R_S^\theta[f] \le 4\frac{\mathfrak{R}(\mathcal{W}_B)}{\theta} + O\left(\frac{\log(1/\delta)}{\sqrt{n}}\right).$$

Moreover, given the assumptions on the data and model class we made, the Rademacher complexity satisfies $\mathfrak{R}(\mathcal{W}) \le B/\sqrt{n}$. What we can learn from this bound is that the relevant quantity for generalization is the ratio of complexity to margin, B/θ.

It's important to understand that margin is a scale-sensitive notion; it only makes sense to talk about it after suitable normalization of the parameter vector. If the norm didn't appear in the bound we could scale up the parameter vector to achieve any margin we want. For linear predictors the Euclidean norm provides a natural and often suitable normalization.

Generalization from algorithms

In the overparameterized regime, there are always an infinite number of models that minimize empirical risk. However, when we run a particular algorithm, the algorithm usually returns only one from this continuum. In this section, we show how directly analyzing algorithmic iteration can itself yield generalization bounds.

One pass optimization of stochastic gradient descent

As we briefly discussed in the optimization chapter, we can interpret the convergence analysis of stochastic gradient descent as directly providing a generalization bound for a particular variant of SGD. Here we give the argument in full detail. Suppose that we choose a loss function that upper bounds the number of mistakes. That is, $loss(\widehat{y}, y) \geq \mathbb{1}\{y\widehat{y} < 0\}$. The hinge loss is one example. Choose the function R to be the risk (not empirical risk!) with respect to this loss function:

$$R[w] = \mathbb{E}[loss(w^T x, y)] .$$

At each iteration, suppose we gain access to an example pair (x_i, y_i) sampled i.i.d. from the a data generating distribution. Then when we run the stochastic gradient method, the iterates are

$$w_{t+1} = w_t - \alpha_t e(w_t^T x_t, y_t) x_t , \quad \text{where} \quad e(z, y) = \frac{\partial loss(z, y)}{\partial z} .$$

Suppose that for all x, $\|x\| \leq D$. Also suppose that $|e(z, y)| \leq C$. Then the SGD convergence theorem tells us that after n steps, starting at $w_0 = 0$ and using an appropriately chosen constant step size, the average of our iterates $\bar{w}_n$ will satisfy

$$\mathbb{P}[\text{sign}(\bar{w}_n^T x) \neq y] \leq \mathbb{E}[R[\bar{w}_n]] \leq R[w_\star] + \frac{CD\|w_\star\|}{\sqrt{n}} .$$

This inequality tells us that we will find a distribution boundary that has low *population* risk after seeing n samples. The population risk itself lets us upper bound the probability that our model makes an error on new data. That is, this inequality is a generalization bound.

We note here that this importantly does not measure our empirical risk. By running stochastic gradient descent, we can find a low-risk model without ever computing the empirical risk.

Let us further assume that the population can be separated with large margin. As we showed when we discussed the perceptron, the margin is equal to the inverse of the norm of the corresponding hyperplane. Suppose we ran the stochastic gradient method using a hinge loss. In this case, $C = 1$, so, letting γ denote the maximum margin, we get the simplified bound

$$\mathbb{P}[\text{sign}(\bar{w}_n^T x) \neq y] \leq \frac{D}{\gamma\sqrt{n}} .$$

Note that the perceptron analysis did not have a step size parameter that depended on the problem instance. But, on the other hand, this analysis of SGD holds regardless of whether the data is separable or whether zero empirical risk is achieved after one pass over the data. The stochastic gradient analysis is more general but generality comes at the cost of a looser bound on the probability of error on new examples.

Uniform stability of stochastic gradient descent

Above, we showed that empirical risk minimization is stable no matter what optimization method we use to solve the objective. One weakness is that the analysis applied to the exact solution of the optimization problem and only applies for strongly convex loss function. In practice, we might only be able to compute an approximate empirical risk minimizer and may be interested in losses that are not strongly convex. Fortunately, we can also show that some optimization methods are stable even if they don't end up computing a minimizer of a strongly convex empirical risk. Specifically, this is true for the stochastic gradient method under suitable assumptions. Below, we state one such result that requires the assumption that the loss function is *smooth*. A continuously differentiable function $f\colon \mathbb{R}^d \to \mathbb{R}$ is β-smooth if $\|\nabla f(y) - \nabla f(x)\| \le \beta\|y - x\|$.

Theorem 6. *Assume a continuously differentiable loss function that is β-smooth and L-Lipschitz on every example and convex. Suppose that we run the stochastic gradient method (SGM) with step sizes $\eta_t \le 2/\beta$ for T steps. Then, we have*

$$\Delta_{\sup}(SGM) \le \frac{2L^2}{n}\sum_{t=1}^{T}\eta_t\,.$$

The theorem allows for SGD to sample the same data points multiple times, as is common practice in machine learning. The stability approach also extends to the non-convex case albeit with a much weaker quantitative bound.

What solutions does stochastic gradient descent favor?

We reviewed empirical evidence that explicit regularization is not necessary for generalization. Researchers therefore believe that a combination of data generating distribution and optimization algorithm perform *implicit regularization*. Implicit regularization describes the tendency of an algorithm to seek out solutions that generalize well on their own on a given a dataset without the need for explicit correction. Since the empirical phenomena we reviewed are all based on gradient methods, it makes sense to study implicit regularization of gradient descent. While a general theory for non-convex problems remains elusive, the situation for linear models is instructive.

Consider again the linear case of gradient descent or stochastic gradient descent:

$$w_{t+1} = w_t - \alpha e_t x_t$$

where e_t is the gradient of the loss at the current prediction. As we showed in the optimization chapter, if we run this algorithm to convergence, we must have that the resulting $\hat{w}$ lies in the span of the data, and that it interpolates the data. These two facts imply that the optimal $\hat{w}$ is the minimum Euclidean norm solution of $Xw = y$. That is, w solves the optimization problem

$$\begin{array}{ll} \text{minimize} & \|w\|^2 \\ \text{subject to} & y_i w^T x_i = 1 . \end{array}$$

Moreover, a closed form solution of this problem is given by

$$\hat{w} = X^T (XX^T)^{-1} y .$$

That is, when we run stochastic gradient descent we converge to a very specific solution. Now what can we say about the generalization properties of this minimum norm interpolating solution?

The key to analyzing the generalization of the minimum norm solution will be a stability-like argument. We aim to control the error of the model trained on the first m data points on the next data point in the sequence, x_{m+1}. To do so, we use a simple identity that follows from linear algebra.

Lemma 3. *Let S be an arbitrary set of $m \geq 2$ data points. Let w_{m-1} and w_m denote the minimum norm solution trained on the first $m-1$ and m points respectively. Then*

$$(1 - y_m \langle w_{m-1}, x_m \rangle)^2 = s_m^2 (\|w_m\|^2 - \|w_{m-1}\|^2) ,$$

where

$$s_m := \operatorname{dist}(\operatorname{span}(x_1, \ldots, x_{m-1}), x_m) .$$

We hold off on proving this lemma and first prove our generalization result with the help of this lemma.

Theorem 7. *Let S_{n+1} denote a set of $n+1$ i.i.d. samples. Let S_j denote the first j samples and w_j denote the solution of minimum norm that interpolates these j points. Let R_j denote the maximum norm of $\|x_i\|$ for $1 \leq i \leq j$. Let (x, y) denote another independent sample from $\mathcal{D}$. Then if $\epsilon_j := \mathbb{E}[(1 - y f_{S_j}(x))^2]$ is a non-increasing sequence, we have*

$$\mathbb{P}[y \langle w_n, x \rangle < 0] \leq \frac{\mathbb{E}[R_j^2 \|w_{n+1}\|^2]}{n} .$$

Proof. Lemma together with the bound $s_i^2 \leq R_{n+1}^2$ yields the inequality

$$\mathbb{E}[(1 - y \langle w_i, x \rangle)^2] \leq (\mathbb{E}[R_{n+1}^2 \|w_{i+1}\|^2] - \mathbb{E}[R_{n+1}^2 \|w_i\|^2]) .$$

Here, we could drop the subscript on x and y on the left-hand side as they are identically distributed to (x_{i+1}, y_{i+1}). Adding these inequalities together gives the bound

$$\frac{1}{n}\sum_{i=1}^{n}\mathbb{E}[(1-yf_{S_i}(x))^2] \leq \frac{\mathbb{E}[R_{n+1}^2\|w_{n+1}\|^2]}{n}.$$

Assuming the sequence is decreasing means that the minimum summand of the previous inequality is $\mathbb{E}[(1-yf_i(x))^2]$. This and Markov's inequality prove the theorem.

□

This proof reveals that the minimum norm solution, the one found by running stochastic gradient descent to convergence, achieves a nearly identical generalization bound as the perceptron, even with the fast $1/n$ rate. Here, nothing is assumed about margin, but instead we assume that the complexity of the interpolating solution does not grow rapidly as we increase the amount of data we collect. This proof combines ideas from stability, optimization, and model complexity to find yet another explanation for why gradient methods find high-quality solutions to machine learning problems.

Proof of Lemma 3

We conclude with the deferred proof of Lemma 3.

Proof. Let $K = XX^T$ denote the kernel matrix for S. Partition K as

$$K = \begin{bmatrix} K_{11} & K_{12} \\ K_{21} & K_{22} \end{bmatrix}$$

where K_{11} is $(m-1)\times(m-1)$ and K_{22} is a scalar equal to $\langle x_m, x_m\rangle$. Similarly, partition the vector of labels y so that $y^{(m-1)}$ denotes the first $m-1$ labels. Under this partitioning,

$$\langle w_{m-1}, x_m\rangle = K_{21}K_{11}^{-1}y^{(m-1)}.$$

Now note that

$$s_m^2 = K_{22} - K_{21}K_{11}^{-1}K_{12}.$$

Next, using the formula for inverting partitioned matrices, we find

$$K^{-1} = \begin{bmatrix} (K_{11} - K_{12}K_{21}K_{22}^{-1})^{-1} & s_m^{-2}K_{11}^{-1}K_{12} \\ s_m^{-2}(K_{11}^{-1}K_{12})^T & s_m^{-2} \end{bmatrix}.$$

By the matrix inversion lemma we have

$$(K_{11} - K_{12}K_{21}K_{22}^{-1})^{-1} = K_{11}^{-1} + s_m^{-2}\left(K_{21}K_{11}^{-1}\right)^T\left(K_{21}K_{11}^{-1}\right).$$

Hence,

$$\begin{aligned}\|w_i\| &= y^T K^{-1} y \\ &= s_m^{-2}(y_m^2 - 2y_m\langle w_{m-1}, x_m\rangle + \langle w_{m-1}, x_m\rangle^2) + y^{(m-1)^T} K_{11}^{-1} y^{(m-1)}.\end{aligned}$$

Rearranging terms proves the lemma.

□

Looking ahead

Despite significant effort and many recent advances, the theory of generalization in overparameterized models still lags behind the empirical phenomenology. What governs generalization remains a matter of debate in the research community.

Existing generalization bounds often do not apply directly to practice by virtue of their assumptions, are quantitatively too weak to apply to heavily overparameterized models, or fail to explain important empirical observations. However, it is not just a lack of quantitative sharpness that limits our understanding of generalization.

Conceptual questions remain open: What is it a successful theory of generalization should do? What are formal success criteria? Even a qualitative theory of generalization that is not quantitatively precise in concrete settings may be useful if it leads to the successful algorithmic interventions. But how do we best evaluate the value of a theory in this context?

Our focus in this chapter was decidedly narrow. We discussed how to relate risk and empirical risk. This perspective can only capture questions that relate performance on a sample to performance on the very same distribution that the sample was drawn from. What is left out are important questions of *extrapolation* from a training environment to testing conditions that differ from training. Overparameterized models that generalize well in the narrow sense can fail dramatically even with small changes in the environment. We will revisit the question of generalization for overparameterized models in our chapter on deep learning.

Chapter notes

The tight characterization of generalization gap in terms of average stability, as well as stability of regularized empirical risk minimization (Theorem 4), is due to Shalev-Shwartz et al.[90] Uniform stability was introduced by Bousquet and Elisseeff.[91] For additional background on VC dimension and Rademacher complexity, see, for example, the text by Shalev-Shwartz and Ben-David.[92]

The double descent figure is from work of Belkin et al.[93] Earlier work pointed out similar empirical risk-complexity relationships.[94] The empirical findings related to the randomization test and the role of regularization are due to Zhang et al.[95]

Theorem 5 is due to Schapire et al.[96] Later work showed theoretically that boosting maximizes margin.[97,98] The margin bound for linear models follows from more general results of Kakade, Sridharan, and Tewari[99] that build on earlier work by Bartlett and Mendelson,[100] as well as work of Koltchinskii and Panchenko.[101] Rademacher complexity bounds for families of neural networks go back to work of Bartlett[102] and remain an active research topic. We will see more on this in our chapter on deep learning.

The uniform stability bound for stochastic gradient descent is due to Hardt, Recht, and Singer.[103] Subsequent work further explores the generalization performance stochastic gradient descent in terms of its stability properties. Theorem 7 and Lemma 3 are due to Liang and Recht.[104]

There has been an explosion of work on generalization and overparameterization in recent years. See, also, recent work exploring how other norms shed light on generalization performance.[105] Our exposition is by no means a representative survey of the broad literature on this topic. There are several ongoing lines of work we did not cover: PAC-Bayes bounds,[106] compression bounds,[107] and arguments about the properties of the optimization landscape.[108] This chapter builds on a chapter by Hardt,[109] but contains several structural changes as well as different results.

Chapter 7

Deep Learning

The past chapters have sketched a path toward predictive modeling: acquire data, construct a set of features that properly represent data in a way such that relevant conditions can be discriminated, pose a convex optimization problem that balances fitting training data to managing model complexity, optimize this problem with a standard solver, and then reason about generalization via the holdout method or cross-validation. In many ways this pipeline suffices for most predictive tasks.

However, this standard practice does have its deficiencies. Feature engineering has many moving pieces, and choices at one part of the pipeline may influence downstream decisions in unpredictable ways. Moreover, different software dependencies may be required to intertwine the various parts of this chain, making the machine learning engineering more fragile. It's additionally possible that more concise feature representations are possible if the pieces can all be tuned together.

Though motivated differently by different people, *deep learning* can be understood as an attempt to "delayer" the abstraction boundaries in the standard machine learning workflow. It enables holistic design of representation and optimization. This delayering comes at the cost of loss of convexity and some theoretical guarantees on optimization and generalization. But, as we will now describe in this chapter, this cost is often quite modest and, on many machine learning problems such as image classification and machine translation, the predictive gains can be dramatic.

Deep learning has been tremendously successful in solving industrial machine learning problems at many tech companies. It is also the top performing approach in most academic prediction tasks in computer vision, speech, and other domains.

The success of deep learning is not just a purely technical matter. Once the industry had embraced deep learning, an unprecedented amount of resources have gone into building and refining high-quality software for practicing deep learning. The open source deep learning ecosystem is vast and evolving quickly. For almost any task, there is already some code available to start from. Companies are actively competing over open source frameworks with convenient high-level syntax and extensive documentation. An honest answer for why practitioners prefer deep learning at this point over other methods is because it simply seems to work better on many problems and there is a lot of quality code available.

We now retrace our path through representation, optimization, and generalization, highlighting what is different for deep learning and what remains the same.

Deep models and feature representation

We discussed in the chapter on representation that template matching, pooling, and nonlinear lifting can all be achieved by affine transformations followed by pointwise nonlinearities. These mappings can be chained together to give a series of new feature vectors:

$$x_{\ell+1} = \phi(A_\ell x_\ell + b_\ell) .$$

Here, ℓ indexes the *layer* of a model. We can chain several layers together to yield a final representation x_L.

As a canonical example, suppose x_1 is a pixel representation of an image. Let's say this representation has size $d_1 \times d_1 \times c_1$, with d_1 counting spatial dimensions and c_1 counting the number of color channels. We could apply c_2 template matching convolutions to this image, resulting in a second layer of size $d_1 \times d_1 \times c_2$. Since we expect convolutions to capture local variation, we can compress the size of this second layer, averaging every 2×2 cell to produce x_2 of size $d_2 \times d_2 \times c_2$, with $d_2 < d_1$ and $c_2 > c_1$. Repeating this procedure several times will yield a representation x_{L-1} that has few spatial dimensions (d_{L-1} is small) but many channel dimensions (c_{L-1} is large). We can then map this penultimate layer through some universal approximator such as a neural network.

A variety of machine learning pipelines can be thought of in this way. The first layer might correspond to edge detectors as in SIFT[110] or HOG.[111] The second layer may look for parts relevant to detection as in a deformable parts model.[112] The insight in deep learning is that we can declare the parameters of each layer A_ℓ and b_ℓ to be *optimization variables*. This way, we do not have to worry about the particular edge or color detectors to use for our problem, but can instead let the collected data dictate the best settings of these features.

This abstraction of "features" as "structure linear maps with tunable parameters" allows for a set of basic building blocks that can be used across a variety of domains.

1. **Fully connected layers.** Fully connected layers are simply unstructured neural networks that we discussed in the representation chapter. For a fixed nonlinear function σ, a fully connected layer maps a vector x to a vector z with coordinates

$$z_i = \sigma\left(\sum_j A_{ij} x_j + b_i\right).$$

While it is popular to chain fully connected layers together to get deep neural networks, there is little established advantage over just using a single layer. Daniely et al. have backed up this empirical observation, showing theoretically that no new approximation power is gained by concatenating fully connected layers.[72] Moreover, as we will discuss below, concatenating many layers together often slows down optimization. As with most things in deep learning, there's nothing saying you can't chain fully connected layers, but we argue that most of the gains in deep learning come from the structured transforms, including the ones we highlight here.

2. **Convolutions**. Convolutions are the most important building block in all of deep learning. We have already discussed the use of convolutions as template matchers that promote spatial invariances. Suppose the input is $d_0 \times d_0 \times c_0$, with the first two components indexing space and the last indexing channels. The parameter A has size $q_0 \times q_0 \times c_0 \times c_1$, where q_0 is usually small (greater than 2 and less than 10). b typically has size c_1. The number of parameters used to define a convolutional layer is dramatically smaller than what would be used in a fully connected layer. The structured linear map of a convolution can be written as

$$z_{a,b,c} = \sigma\left(\sum_{i,j,k} A_{i,j,k,c} x_{a-i,b-j,k} + b_c\right).$$

3. **Recurrent structures.** Recurrent structures let us capture repeatable stationary patters in time or space. Suppose we expect stationarity in time. In this case, we expect each layer to represent a state of the system, and the next time step should be a static function of the previous one:

$$x_{t+1} = f(x_t).$$

When we write f as a neural network, this is called a *recurrent neural network*. Recurrent neural networks *share weights* insofar as the f does not change from one time step to the next.

4. **Attention mechanisms.** Attention mechanisms have proven powerful tools in natural language processing for collections of vectors with dependencies that are not necessarily sequential. Suppose our layer is a list of m vectors of dimension d. That is, x has shape $d \times m$. An attention layer will have two matrices U and V, one to operate on the feature dimensions and one to operate on the sequential dimensions. The transformation takes form

$$z_{a,b} = \sigma\left(\sum_{i,j} U_{a,i} x_{ij} V_{b,j}\right).$$

Just as was the case with convolutions, this structured map can have fewer dimensions than a fully connected layer, and can also respect a separation of the feature dimensions from the sequential dimensions in the data matrix x.

Optimization of deep nets

Once we have settled on a feature representation, typically called *model architecture* in this context, we now need to solve empirical risk minimization. Let's group all of the parameters of the layers into a single large array of weights, w. We denote the map from x to prediction with weights w as $f(x; w)$. At an abstract level, empirical risk minimization amounts to minimizing

$$R_S[w] = \frac{1}{n} \sum_{i=1}^{n} loss(f(x_i; w), y_i).$$

This is a nonconvex optimization problem, but we can still run gradient methods to try to find minimizers. Our main concerns from an optimization perspective are whether we run into local optima and how can we compute gradient steps efficiently.

We will address gradient computation through a discussion of automatic differentiation. With regard to global optimization, there are unfortunately computational complexity results proving efficient optimization of arbitrary neural networks is intractable. Even neural nets with a single neuron can have exponentially many local minimizers,[113] and finding the minimum of a simple two-layer neural network is NP-hard.[114,115] We cannot expect a perfectly clean mathematical theory guiding our design and implementation of neural net optimization algorithms.

However, these theoretical results are about the *worst case*. In practice, optimization of neural nets is often easy. If the loss is bounded below by zero, any model with zero loss is a global minimizer. As we discussed in the generalization chapter, one can quickly find a global optimum of a state-of-the-art neural net by cloning a GitHub repo and turning off the various regularization schemes. In this section, we aim to provide some insights about the disconnect between computational complexity in the worst case and the results achieved in practice. We provide some partial insights as to why neural net optimization is doable by studying the convergence of the predictions and how this convergence can be aided by overparameterization.

Convergence of predictions in nonlinear models

Consider the special case where we aim to minimize the square loss. Let $\widehat{y}_t$ denote the vector of predictions $\left(f(x_i;w_t)\right)_{i=1}^n \in \mathbb{R}^n$. Gradient descent follows the iterations

$$w_{t+1} = w_t - \alpha \mathsf{D}_w f(x;w_t)(\widehat{y}_t - y)$$

where $\mathsf{D}_w f(x;w_t)$ denotes the $d \times n$ Jacobian matrix of the predictions $\widehat{y}_t$. Reasoning about convergence of the weights is difficult, but we showed in the optimization chapter that we could reason about convergence of the predictions:

$$\widehat{y}_{t+1} - y = (I - \alpha \mathsf{D}_w f(x;w_t)^T \mathsf{D}_w f(x;w_t))(\widehat{y}_t - y) + \alpha \epsilon_t\,.$$

Here,

$$\epsilon_t = \frac{\alpha}{2}\Lambda\|\widehat{y}_t - y\|^2$$

and Λ bounds the curvature of the f. If $\epsilon_t = 0$ is sufficiently small the predictions will converge to the training labels.

Hence, under some assumptions, nonconvexity does not stand in the way of convergence to an empirical risk minimizer. Moreover, control of the Jacobians $\mathsf{D}_w f$ can accelerate convergence. We can derive some reasonable ground rules on how to keep $\mathsf{D}_w f$ well conditioned by unpacking how we compute gradients of compositions of functions.

Automatic differentiation

For linear models it's not hard to calculate the gradient with respect to the model parameters analytically and to implement the analytical expression efficiently. The situation changes once we stack many operations on top of each other. Even though in principle we can calculate gradients using the chain rule, this gets tedious and error-prone very quickly. Moreover, the straightforward way of implementing the chain rule is computationally inefficient. What's worse is that any change to the model architecture would require us to redo our calculation and update its implementation.

Fortunately, the field of *automatic differentiation* has the tools to avoid all of these problems. At a high level, automatic differentiation provides efficient algorithms to compute gradients of any function that we can write as a composition of differentiable building blocks. Though automatic differentiation has existed since the 1960s, the success of deep learning has led to numerous free, well-engineered, and efficient automatic differentiation packages. Moreover, the dynamic programming algorithm behind these methods is quite instructive as it gives us some insights into how to best engineer model architectures with desirable gradients.

Automatic differentiation serves two useful purposes in deep learning. First, it lowers the barrier to entry, allowing practitioners to think more about modeling than about the particulars of calculus and numerical optimization. Second, a standard automatic differentiation algorithm helps us write down parseable formulas for the gradients of our objective function so we can reason about what structures encourage faster optimization.

To define this dynamic programming algorithm, called *backpropagation*, it's helpful to move up a level of abstraction before making the matter more concrete again. After all, the idea is completely general and not specific to deep models. Specifically, we consider an abstract computation that proceeds in L steps starting from an input $z^{(0)}$ and produces an output $z^{(L)}$ with $L-1$ intermediate steps $z^{(1)}, \dots, z^{(L-1)}$:

$$\begin{aligned} \text{input} \quad & z^{(0)} \\ & z^{(1)} = f_1(z^{(0)}) \\ & \quad \vdots \\ & z^{(L-1)} = f_{L-1}(z^{(L-2)}) \\ \text{output} \quad & z^{(L)} = f_L(z^{(L-1)}) \,. \end{aligned}$$

We assume that each *layer* $z^{(i)}$ is a real-valued vector and that each function f_i maps a real-valued vector of some dimension to another dimension. Recall that for a function $f\colon \mathbb{R}^n \to \mathbb{R}^m$, the Jacobian matrix $\mathsf{D}(w)$ is the $n \times m$ matrix of first-order partial derivatives evaluated at the point w. When $m = 1$ the Jacobian matrix coincides with the transpose of the gradient.

In the context of backpropagation, it makes sense to be a bit more explicit in our notation. Specifically, we will denote the Jacobian of a function f with respect to a variable x evaluated at point w by $\mathsf{D}_x f(w)$.

The backpropagation algorithm is a computationally efficient way of computing the partial derivatives of the output $z^{(L)}$ with respect to the input $z^{(0)}$ evaluated at a given parameter vector w, that is, the Jacobian $\mathsf{D}_{z^{(0)}} z^{(L)}(w)$. Along the way, it computes the Jacobians for any of the intermediate layers as well.

Backpropagation

- Input: parameters w.
- Forward pass:
  - Set $v_0 = w$.
  - For $i = 1, \ldots, L$:
    * Store v_{i-1} and compute $v_i = f_i(v_{i-1})$
- Backward pass:
  - Set $\Lambda_L = \mathsf{D}_{z_L} z_L(v_L) = I$.
  - For $i = L, \ldots, 1$:
    * Set $\Lambda_{i-1} = \Lambda_i \mathsf{D}_{z^{(i-1)}} z^{(i)}(v_{i-1})$.
- Output Λ_0.

First note that backpropagation runs in time $O(LC)$ where C is the computational cost of performing an operation at one step. On the forward pass, this cost corresponds to function evaluation. On the backward pass, it requires computing the partial derivatives of $z^{(i)}$ with respect to $z^{(i-1)}$. The computational cost depends on what the operation is. Typically, the ith step of the backward pass has the same computational cost as the corresponding ith step of the forward pass up to constant factors. What is important is that computing these partial derivatives is an entirely *local* operation. It only requires the partial derivatives of a function with respect to its input evaluated at the value v_{i-1} that we computed in the forward pass. This observation is key to all fast implementations of backpropagation. Each operation in our computation only has to implement function evaluation, its first-order derivatives, and store an array of values computed in the forward pass. There is nothing else we need to know about the computation that comes before or after.

The main correctness property of the algorithm is that the final output Λ_0 equals the partial derivative of the output layer with respect to the input layer evaluated at the input w.

Proposition 8. ***Correctness of backpropagation.***

$$\Lambda_0 = \mathsf{D}_{z^{(0)}} z^{(L)}(w) \,.$$

The claim directly follows by induction from the following lemma, which states that if we have the correct partial derivatives at step i of the backward pass, then we also get them at step $i - 1$.

Lemma 4. ***Backpropagation invariant.***

$$\Lambda_i = \mathsf{D}_{z^{(i)}} z^{(L)}(v_i) \quad \Longrightarrow \quad \Lambda_{i-1} = \mathsf{D}_{z^{(i-1)}} z^{(L)}(v_{i-1}) \,.$$

Proof. Assume that the premise holds. Then, we have

$$\begin{aligned}\Lambda_{i-1} &= \Lambda_i \mathsf{D}_{z^{(i-1)}} z^{(i)}(v_{i-1}) \\ &= \mathsf{D}_{z^{(i)}} z^{(L)}(v_i) \mathsf{D}_{z^{(i-1)}} z^{(i)}(v_{i-1}) \\ &= \mathsf{D}_{z^{(i-1)}} z^{(L)}(v_{i-1})\end{aligned}$$

The last identity is the multivariate chain rule.

□

To aid the intuition, it can be helpful to write the multivariate chain rule informally in the familiar Leibniz notation:

$$\frac{\partial z^{(L)}}{\partial z^{(i)}} \frac{\partial z^{(i)}}{\partial z^{(i-1)}} = \frac{\partial z^{(L)}}{\partial z^{(i-1)}}.$$

A worked out example

The backpropagation algorithm works in great generality. It produces partial derivatives for any variable appearing in any layer $z^{(i)}$. So, if we want partial derivatives with respect to some parameters, we only need to make sure they appear on one of the layers. This abstraction is so general that we can easily express all sorts of deep architectures and the associated objective functions with it.

But let's make that more concrete and get a feeling for the mechanics of backpropagation in the two cases that are most common: a nonlinear transformation applied coordinate-wise, and a linear transformation.

Suppose at some point in our computation we apply the *rectified linear unit* $\mathrm{ReLU} = \max\{u, 0\}$ coordinate-wise to the vector u. ReLU units remain one of the most common nonlinearities in deep neural networks. To implement the backward pass, all we need to be able to do is to compute the partial derivatives of $\mathrm{ReLU}(u)$ with respect to u. It's clear that when $u_i > 0$, the derivative is 1. When $u_i < 0$, the derivative is 0. The derivative is not defined at $u_i = 0$, but we choose to ignore this issue by setting it to be 0. The resulting Jacobian matrix is a diagonal matrix $D(u)$ that has 1 in all coordinates corresponding to positive coordinates of the input vector, and has 0 elsewhere.

- Forward pass:
  - Input: u.
  - Store u and compute the value $v = \mathrm{ReLU}(u)$.
- Backward pass:
  - Input: Jacobian matrix Λ.
  - Output $\Lambda D(u)$.

If we were to swap out the rectified linear unit for some other coordinate-wise nonlinearity, we'd still get a diagonal matrix. What changes are the coefficients along the diagonal.

Now, let's consider an affine transformation of the input $v = Au + b$. The Jacobian of an affine transformation is simply the matrix A itself. Hence, the backward pass is simple:

- Backward pass:
  - Input: Jacobian matrix Λ.
  - Output: ΛA.

We can now easily chain these together. Suppose we have a typical ReLU network that strings together L linear transformations with ReLU operations in between:

$$f(x) = A_L \mathrm{ReLU}(A_{L-1}\mathrm{ReLU}(\cdots \mathrm{ReLU}(A_1 x)))\,.$$

The Jacobian of this chained operation with respect to the input x is given by a chain of linear operations:

$$A_L D_{L-1} A_{L-1} \cdots D_1 A_1\,.$$

Each matrix D_i is a diagonal matrix that zeroes out some coordinates and leaves others untouched. Now, in a deep model architecture the weights of the linear transformation are typically trainable. That means we really want to be able to compute the gradients with respect to, say, the coefficients of the matrix A_i.

Fortunately, the same reasoning applies. All we need to do is to make A_i part of the input to the matrix-vector multiplication node and backpropagation will automatically produce derivatives for it. To illustrate this point, suppose we want to compute the partial derivatives with respect to, say, the jth column of A_i. Let $u = \mathrm{ReLU}(A_{i-1}\cdots(A_1 x))$ be the vector that we multiply A_i with. We know from the argument above that the Jacobian of the output with respect to the vector $A_i u$ is given by the linear map $B = A_L D_{L-1} A_{L-1} \cdots D_i$. To get the Jacobian with respect to the jth column of A_i we therefore only need to find the partial derivative of $A_i u$ with respect to the jth column of A_i. We can verify by taking derivatives that this equals $u_j I$. Hence, the partial derivatives of $f(x)$ with respect to the jth column of A_i are given by Bu_j.

Let's add in the final ingredient, a loss function. Suppose A_L maps into one dimension and consider the squared loss $\frac{1}{2}(f(x) - y)^2$. The only thing this loss function does is to scale our Jacobian by a factor $f(x) - y$. In other words, the partial derivatives of the loss with respect to the jth columns of the weight matrix A_i are given by $(f(x) - y)Bu_j$.

As usual with derivatives, we can interpret this quantity as the *sensitivity* of our loss to changes in the model parameters. This interpretation will be helpful next.

Vanishing gradients

The previous discussion revealed that the gradients of deep models are produced by chains of linear operations. Generally speaking, chains of linear operations tend to either blow up the input exponentially with depth, or shrink it, depending on the singular values of the matrices. It is helpful to keep in mind the simple case of powering the same symmetric real matrix A a number of times. For almost all vectors u, the norm $\|A^L u\|$ grows as $\Theta\left(\lambda_1(A)^L\right)$ asymptotically with L. Hence it vanishes exponentially quickly if $\lambda_1(A) < 1$ and it grows exponentially quickly if $\lambda_1(A) > 1$.

When it comes to gradients, these two cases correspond to the *vanishing gradients problem* and the *exploding gradients problem*, respectively. Both result in a failure case for gradient-based optimization methods. Huge gradients are numerically unstable. Tiny gradients preclude progress and stall any gradient-based optimization method. We can always avoid one of the problems by scaling the weights, but avoiding both can be delicate.

Vanishing and exploding gradients are not the fault of an optimization method. They are a property of the model architecture. As a result, to avoid them we need to engineer our model architectures carefully. Many architectural innovations in deep learning aim to address this problem. We will discuss two of them. The first is the residual connection, and the second is layer normalizations.

Residual connections

The basic idea behind *residual networks* is to make each layer close to the identity map. Traditional building blocks of neural networks typically looked like two affine transformations A, B, with a nonlinearity in the middle:

$$f(x) = B(\mathrm{ReLU}(A(x)))$$

Residual networks modify these building blocks by adding the input x back to the output:

$$h(x) = x + B(\mathrm{ReLU}(A(x)))\,.$$

In cases where the output of C differs in its dimension from x, practitioners use different padding or truncation schemes to match dimensions. Thinking about the computation graph, we create a connection from the input to the output of the building block that *skips* the transformation. Such connections are therefore called *skip connections*.

This seemingly innocuous change was hugely successful. Residual networks took the computer vision community by storm, after they achieved leading performance in numerous benchmarks upon their release in 2015. These networks seemed to avoid the vanishing gradients better than prior architectures and allowed for model depths not seen before.

Let's begin to get some intuition for why residual layers are reasonable by thinking about what they do to the gradient computation. Let $J = \mathsf{D}_x f(x)$ be the Jacobian of the function f with respect to its input. We can verify that the Jacobian of the residual block looks like

$$J' = \mathsf{D}_x h(x) = \mathsf{D}_x I(x) + \mathsf{D}_x f(x) = I + J\,.$$

In other words, the Jacobian of a residual block is the Jacobian of a regular block plus the identity. This means that if we scale down the weights of the regular block, the Jacobian J' approaches the identity in the sense that all its singular values are between $1-\epsilon$ and $1+\epsilon$. We can think of such a transformation as locally well conditioned. It neither blows up nor shrinks down the gradient much. Since the full Jacobian of a deep residual network will be a product of such matrices, our reasoning suggests that suitably scaled residual networks have well-conditioned Jacobians, and hence, as we discussed above, predictions should converge rapidly.

Our analyses of non-convex ERMs thus far have described scenarios where we could prove that the *predictions* converge to the training labels. Residual networks are interesting as we can construct cases where the *weights* converge to a unique optimal solution. This perhaps gives even further motivation for their use.

Let's consider the simple case of where the activation function is the identity. The resulting residual blocks are linear transformations of the form $I + A$. We can chain them as $A = (I + A_L) \cdots (I + A_1)$. Such networks are no longer universal approximators, but they are non-convex parameterizations of linear functions. We can turn this parameterization into an optimization problem by solving a least squares regression problem in this residual parameterization:

$$\text{minimize}_{A_1,\dots,A_L} \quad \mathbb{E}[\tfrac{1}{2}\|A(X) - Y\|^2]\,.$$

Assume X is a random vector with covariance matrix I and $Y = B(X) + G$ where B is a linear transformation and G is centered random Gaussian noise. A standard trick shows that up to an additive constant the objective function is equal to

$$f(A_1, \dots, A_L) = \frac{1}{2}\|A - B\|_F^2\,.$$

What can we say about the gradients of this function? We can verify that the Jacobian of f with respect to A_i equals

$$\mathsf{D}_{A_i} f(A_1, \dots, A_L) = P^T E Q^T$$

where $P = (I + A_L) \cdots (I + A_{i+1})$ and $Q = (I + A_{i-1}) \cdots (I + A_1)$.

Note that when P and Q are non-singular, then the gradient can only vanish at $E = 0$. But that means we're at the global minimum of the objective function f. We can ensure that P and Q are non-singular by making the largest singular value of each A_i to be less than $1/L$.

The property we find here is that the gradient vanishes only at the optimum. This is implied by convexity, but it does not imply convexity. Indeed, the objective function above is not convex. However, this weaker property is enough to ensure that gradient-based methods do not get stuck except at the optimum. In particular, the objective has no saddle points. This desirable property does not hold for the standard parameterization $A = A_L \cdots A_1$ and so it speaks to the benefit of the residual parameterization.

Normalization

Consider a feature vector x. We can partition this vector into subsets so that

$$x = \begin{bmatrix} x_1 & \cdots & x_P \end{bmatrix},$$

and *normalize* each subset to yield a vector with partitions

$$x_i' = 1/s_i(x_i - \mu_i)$$

where μ_i is the mean of the components of x_i and s_i is the standard deviation.

Such normalization schemes have proven powerful for accelerating the convergence of stochastic gradient descent. In particular, it is clear why such an operation can improve the conditioning of the Jacobian. Consider the simple linear case where $\hat{y} = Xw$. Then $\mathsf{D}(w) = X$. If each row of X has a large mean, i.e., $X \approx X_0 + c\mathbb{1}\mathbb{1}^T$, then X may be ill conditioned, as the rank-one term will dominate the first singular value, and the remaining singular values may be small. Removing the mean improves the condition number. Rescaling by the variance may improve the condition number, but also has the benefit of avoiding numerical scaling issues, forcing each layer in a neural network to be on the same scale.

Such whole vector operations are expensive. Normalization in deep learning chooses parts of the vector to normalize that can be computed quickly. Batch Normalization normalizes along the data dimension in batches of data used as stochastic gradient minibatches.[116] Group Normalization generalizes this notion to arbitrary partitioning of the data, encompassing a variety of normalization proposals.[117] The best normalization scheme for a particular task is problem dependent, but there is a great deal of flexibility in how one partitions features for normalization.

Generalization in deep learning

While our understanding of optimization of deep neural networks has made significant progress, our understanding of generalization is considerably less settled. In the previous chapter, we highlighted four paths toward generalization: stability, capacity, margin, optimization. It's plausible deep neural networks have elements of all four of these core components. The evidence is not as cut and dry as it is for linear models, but some mathematical progress has been made to understand how deep learning leverages classic foundations of generalization. In this section we review the partial evidence gathered so far.

In the next chapter we will extend this discussion by taking a closer look at the role that data and community practices play in the study of generalization for deep learning.

Algorithmic stability of deep neural networks

We discussed how stochastic gradient descent trained on convex models was algorithmically stable. The results we saw in Chapter 6 somewhat extend to the non-convex case. However, results that go through uniform stability ultimately cannot explain the generalization performance of training large deep models. The reason is that uniform stability does not depend on the data generating distribution. In fact, uniform stability is invariant under changes to the labeling of the data. But we saw in Chapter 6 that we can make the generalization gap whatever we want by randomizing the labels in the training set.

Nevertheless, it's possible that a weaker notion of stability that is sensitive to the data would work. In fact, it does appear to be the case in experiment that stochastic gradient descent is relatively stable in concrete instances. To measure stability empirically we can look at the Euclidean distance between the parameters of two identical models trained on the datasets that differ only by a single example. If these are close for many independent runs, then the algorithm appears to be stable. Note that parameter distance is a *stronger* notion than stability of the loss.

The plot below displays the parameter distance for the venerable AlexNet model trained on the ImageNet benchmark. We observe that the parameter distance grows sub-linearly even though our theoretical analysis is unable to prove this to be true.

Capacity of deep neural networks

Many researchers have attempted to compute notions like VC-dimension or Rademacher complexity of neural networks. The earliest work by Baum and Haussler bounded the VC-dimension of small neural networks and showed that as long these networks could be optimized well in the *underparameterized*

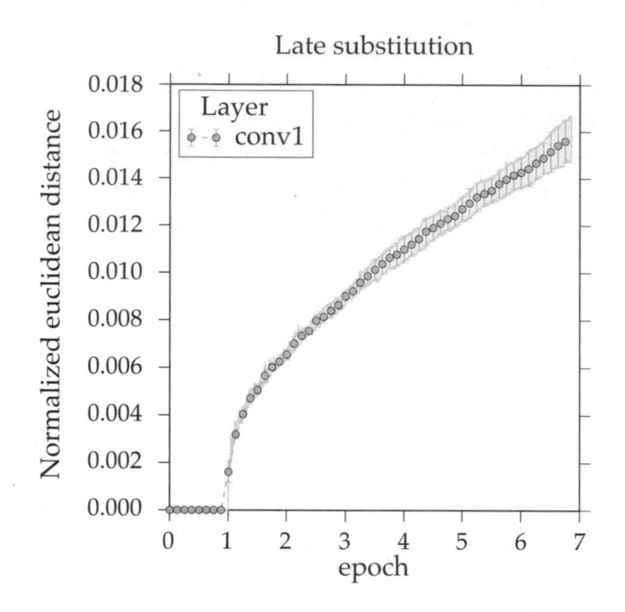

Figure 7.1: Parameter divergence on AlexNet trained on ImageNet. The two models differ only in a single example.

regime, the neural networks would generalize.[118] Later, seminal work by Bartlett showed that the size of the weights was more important than the number of weights in terms of understanding generalization capabilities of neural networks.[119] These bounds were later sharpened using Rademacher complexity arguments.[100]

Margin bounds for deep neural networks

The margin theory for linear models conceptually extends to neural networks. The definition of margin is unchanged. It simply quantifies how close the network is to making an incorrect prediction. What changes is that for multi-layer neural networks the choice of a suitable norm is substantially more delicate.

To see why, a little bit of notation is necessary. We consider multi-layer neural networks specified by a composition of L layers. Each layer is a linear transformation of the input, followed by a coordinate-wise nonlinear map:

$$\text{Input}\, x \to Ax \to \sigma(Ax)\,.$$

The linear transformation has trainable parameters, while the nonlinear map does not. For notational simplicity, we assume we have the same nonlinearity σ at each layer, scaled so that the map is 1-Lipschitz. For example, the popular coordinate-wise ReLU $\max\{x, 0\}$ operation satisfies this assumption.

Given L weight matrices $\mathcal{A} = (A_1, \ldots, A_L)$ let $f_{\mathcal{A}} \colon \mathbb{R}^d \to \mathbb{R}^k$ denote the function computed by the corresponding network:

$$f_{\mathcal{A}}(x) := A_L \sigma(A_{L-1} \cdots \sigma(A_1 x) \cdots)) .$$

The network output $F_{\mathcal{A}}(x) \in \mathbb{R}^k$ is converted to a class label in $\{1, \ldots, k\}$ by taking the arg max over components, with an arbitrary rule for breaking ties. We assume $d \geq k$ only for notational convenience.

Our goal now is to define a complexity measure of the neural network that will allow us to prove a margin bound. Recall that margins are meaningless without a suitable normalization of the network. Convince yourself that the Euclidean norm can no longer work for multi-layer ReLU networks. After all we can scale the linear transformation on one later by a constant c and a subsequent layer by a constant $1/c$. Since the ReLU nonlinearity is piecewise linear, this transformation changes the Euclidean norm of the weights arbitrarily without changing the function that the network computes.

There's much ongoing work about what good norms are for deep neural networks. We will introduce one possible choice here.

Let $\| \cdot \|_{\mathrm{op}}$ denote the spectral norm. Also, let $\|A\|_{2,1}$ be the matrix norm where we apply the ℓ_2-norm to each column of the matrix and then take the ℓ_1-norm of the resulting vector.

The *spectral complexity* $R_{\mathcal{A}}$ of a network $F_{\mathcal{A}}$ with weights $\mathcal{A}$ is the defined as

$$R_{\mathcal{A}} := \left(\prod_{i=1}^{L} \|A_i\|_{\mathrm{op}} \right) \left(\sum_{i=1}^{L} \left(\frac{\|A_i^\top - M_i^\top\|_{2,1}}{\|A_i\|_{\mathrm{op}}} \right)^{2/3} \right)^{3/2} . \tag{7.1}$$

Here, the matrices $M_1, \ldots, M_L$ are free parameters that we can choose to minimize the bound. Random matrices tend to be good choices.

The following theorem provides a generalization bound for neural networks with fixed nonlinearities and weight matrices $\mathcal{A}$ of bounded spectral complexity $R_{\mathcal{A}}$.

Theorem 8. *Assume data $(x_1, y_1), \ldots, (x_n, y_n)$ are drawn i.i.d. from any probability distribution over $\mathbb{R}^d \times \{1, \ldots, k\}$. With probability at least $1 - \delta$, for every margin $\theta > 0$ and every network $f_{\mathcal{A}} \colon \mathbb{R}^d \to \mathbb{R}^k$,*

$$R[f_{\mathcal{A}}] - R_S^\theta[f_{\mathcal{A}}] \leq \widetilde{\mathcal{O}} \left(\frac{R_{\mathcal{A}} \sqrt{\sum_i \|x_i\|_2^2 \ln(d)}}{\theta n} + \sqrt{\frac{\ln(1/\delta)}{n}} \right) ,$$

where $R_S^\theta[f] \leq n^{-1} \sum_i \mathbf{1} \left[f(x_i)_{y_i} \leq \theta + \max_{j \neq y_i} f(x_i)_j \right]$.

The proof of the theorem involves Rademacher complexity and so-called data-dependent covering arguments. Although it can be shown empirically that the above complexity measure $R_{\mathcal{A}}$ is somewhat correlated with generalization performance in some cases, there is no reason to believe that it is the "right" complexity measure. The bound has other undesirable properties, such as an exponential dependence on the depth of the network as well as an implicit dependence on the size of the network.

Implicit regularization by stochastic gradient descent in deep learning

There have been recent attempts to understand the dynamics of stochastic gradient descent on deep neural networks. Using an argument similar to the one we used to reason about convergence of the *predictions* of deep neural networks, Jacot et al. used a differential equations argument to understand to which *function* stochastic gradient converged.[73]

Here we can sketch the rough details of their argument. Recall our expression for the dynamics of the predictions in gradient descent:

$$\widehat{y}_{t+1} - y = (I - \alpha \mathsf{D}(w_t)^T \mathsf{D}(w_t))(\widehat{y}_t - y) + \alpha \epsilon_t \,.$$

If α is very small, we can further approximate this as

$$\widehat{y}_{t+1} - y = (I - \alpha \mathsf{D}(w_0)^T \mathsf{D}(w_0))(\widehat{y}_t - y) + \alpha \epsilon'_t \,.$$

Now ϵ'_t captures both the curvature of the function representation and the deviation of the weights from their initial condition. Note that in this case, $\mathsf{D}(w_0)$ is a constant for all time. The matrix $\mathsf{D}(w_0)^T\mathsf{D}(w_0)$ is $n \times n$, positive semidefinite, and only depends on the data and the initial setting of the weights. When the weights are random, this is a kernel induced by random features. The expected value of this random feature embedding is called a *Neural Tangent Kernel*.

$$k(x_1, x_2) = \mathop{\mathbb{E}}_{w_0}\left[\langle \nabla_w f(x_1; w_0), \nabla_w f(x_2; w_0)\rangle\right] \,.$$

Using a limit where α tends to zero, Jacot et. al argue that a deep neural net will find a minimum norm solution in the RKHS of the Neural Tangent Kernel. This argument was made non-asymptotic by Heckel and Soltanolkotabi.[120] In the generalization chapter, we showed that the minimum norm solution of in RKHS generalized with a rate of $O(1/n)$. Hence, this argument suggests a similar rate of generalization may occur in deep learning, provided the norm of the minimum norm solution does not grow too rapidly with the number of data points and the true function optimized by stochastic gradient descent isn't too dissimilar from the Neural Tangent Kernel limit. We note that this argument is qualitative, and there remains work to be done to make these arguments fully rigorous.

Perhaps the most glaring issue with NTK arguments is that they do not reflect practice. Models trained with Neural Tangent Kernels do not match the predictive performance of the corresponding neural network. Moreover, simpler kernels inspired by these networks can outperform Neural Tangent Kernels.[75] There is a significant gap between the theory and practice here, but the research is new and remains active and this gap may be narrowed in the coming years.

Chapter notes

Deep learning is at this point a vast field with tens of thousands of papers. We don't attempt to survey or systematize this vast literature. It's worth emphasizing though the importance of learning about the area by experimenting with available code on publicly available datasets. The next chapter will cover datasets and benchmarks in detail.

Apart from the development of benchmark datasets, one of the most important advances in deep learning over the last decade has been the development of high-quality, open source software. This software makes it easier than ever to prototype deep neural network models. One such open source project is PyTorch (`pytorch.org`), which we recommend for the researcher interested in experimenting with deep neural networks. The best way to begin to understand some of the nuances of architecture selection is to find preexisting code and understand how it is composed. We recommend the tutorial by David Page that demonstrates how the many different pieces fit together in a deep learning pipeline.[121]

For a more in-depth understanding of automatic differentiation, we recommend Griewank and Walther's *Evaluating Derivatives*.[122] This text works through a variety of unexpected scenarios—such as implicitly defined functions—where gradients can be computed algorithmically. Jax is open source automatic differentiation package that incorporates many of these techniques, and is useful for any application that could be assisted by automatic differentiation.[123]

The AlexNet architecture was introduced by Krizhevsky, Sutskever, and Hinton in 2012.[124] Achieving the best known performance on the ImageNet benchmark at the time, it was pivotal in launching the most recent wave of research on deep learning.

Residual networks were introduced by He, Zhang, Ren, and Sun in 2016.[86] The observation about linear residual networks is due to Hardt and Ma.[125]

Theorem 8 is due to Bartlett, Foster, and Telgarsky.[126] The dimension dependence of the theorem can be removed.[127]

Chapter 8

Datasets

It's become commonplace to point out that machine learning models are only as good as the data they're trained on. The old slogan "garbage in, garbage out" no doubt applies to machine learning practice, as does the related catchphrase "bias in, bias out." Yet, these aphorisms still understate—and somewhat misrepresent—the significance of data for machine learning.

It's not only the output of a learning algorithm that may suffer with poor input data. A dataset serves many other vital functions in the machine learning ecosystem. The dataset itself is an integral part of the problem formulation. It implicitly sorts out and operationalizes what the problem is that practitioners end up solving. Datasets have also shaped the course of entire scientific communities in their capacity to measure and benchmark progress, support competitions, and interface between researchers in academia and practitioners in industry.

If so much hinges on data in machine learning, it might come as a surprise that there is no simple answer to the question of what makes data good for what purpose. The collection of data for machine learning applications has not followed any established theoretical framework, certainly not one that was recognized a priori.

In this chapter, we take a closer look at popular datasets in the field of machine learning and the benchmarks that they support. We trace out the history of benchmarks and work out the implicit scientific methodology behind machine learning benchmarks. We limit the scope of this chapter in some important ways. Our focus will be largely on publicly available datasets that support training and testing purposes in machine learning research and applications. Primarily, we critically examine the train-and-test paradigm machine learning practitioners take for granted today.

The scientific basis of machine learning benchmarks

Methodologically, much of modern machine learning practice rests on a variant of *trial and error*, which we call the *train-test paradigm*. Practitioners repeatedly build models using any number of heuristics and test their performance to see what works. Anything goes as far as training is concerned, subject only to computational constraints, so long as the performance looks good in testing. Trial and error is sound so long as the testing protocol is robust enough to absorb the pressure placed on it. We will examine to what extent this is the case in machine learning.

From a theoretical perspective, the best way to test the performance of a predictor f is to collect a sufficiently large fresh dataset S and to compute the empirical risk $R_S[f]$. We already learned that the empirical risk in this case is an unbiased estimate of the risk of the predictor. For a bounded loss function and a test set of size n, an appeal to Hoeffding's inequality proves the generalization gap to be no worse than $O(1/\sqrt{n})$. We can go a step further and observe that if we take union bound over k fixed predictors, our fresh sample will simultaneously provide good estimates for all k predictors up to a maximum error of $O(\sqrt{\log(k)/n})$. In fact, we can apply any of the mathematical tools we saw in the Generalization chapter so long as the sample S really is a fresh sample with respect to the set of models we want to evaluate.

Data collection, however, is a difficult and costly task. In most applications, practitioners cannot sample fresh data for each model they would like to try out. A different practice has therefore become the de facto standard. Practitioners split their dataset into typically two parts, a *training set* used for training a model, and a *test set* used for evaluating its performance. Sometimes practitioners divide their data into multiple splits, e.g., training, validation, and test sets. However, for our discussion here that won't be necessary. Often the split is determined when the dataset is created. Datasets used for benchmarks in particular have one fixed split persistent throughout time. A number of variations on this theme go under the name *holdout method*.

Machine learning competitions have adopted the same format. The company Kaggle, for example, has organized hundreds of competitions since it was founded. In a competition, a holdout set is kept secret and is used to rank participants on a public leaderboard as the competition unfolds. In the end, the final winner is whoever scores highest on a separate secret test set not used to that point.

In all applications of the holdout method the hope is that the test set will serve as a fresh sample that provides good risk estimates for all the models. The central problem is that practitioners don't just use the test data once only to retire it immediately thereafter. The test data are used incrementally for building one model at a time while incorporating feedback received previously from the test data. This leads to the fear that eventually models begin to *overfit* to the test data.

Duda and Hart summarize the problem aptly in their 1973 textbook:

> In the early work on pattern recognition, when experiments were often done with very small numbers of samples, the same data were often used for designing and testing the classifier. This mistake is frequently referred to as "testing on the training data." A related but less obvious problem arises when a classifier undergoes a long series of refinements guided by the results of repeated testing on the same data. This form of "training on the testing data" often escapes attention until new test samples are obtained.[128]

Nearly half a century later, Hastie, Tibshirani, and Friedman still caution in the 2017 edition of their influential textbook:

> Ideally, the test set should be kept in a "vault," and be brought out only at the end of the data analysis. Suppose instead that we use the test-set repeatedly, choosing the model with smallest test-set error. Then the test set error of the final chosen model will underestimate the true test error, sometimes substantially.[129]

Indeed, reuse of test data—on the face of it—invalidates the statistical guarantees of the holdout method. The predictors created with knowledge about prior test-set evaluations are no longer independent of the test data. In other words, the sample isn't fresh anymore. While the suggestion to keep the test data in a "vault" is safe, it couldn't be further from the reality of modern practice. Popular test datasets often see tens of thousands of evaluations.

We could try to salvage the situation by relying on uniform convergence. If all models we try out have sufficiently small complexity in some formal sense, such as VC-dimension, we could use the tools from the Generalization chapter to negotiate some sort of a bound. However, the whole point of the train-test paradigm is not to constrain the complexity of the models a priori, but rather to let the practitioner experiment freely. Moreover, if we had an actionable theoretical generalization guarantee to begin with, there would hardly be any need for the holdout method, whose purpose is to provide an empirical estimate where theoretical guarantees are lacking.

Before we discuss the "training on the testing data" problem any further, it's helpful to get a better sense of concrete machine learning benchmarks, their histories, and their impact within the community.

A tour of datasets in different domains

The creation of datasets in machine learning does not follow a clear theoretical framework. Datasets aren't collected to test a specific scientific hypothesis. In fact, we will see that there are many different roles that data plays in machine learning. As a result, it makes sense to start by looking at a few influential datasets from different domains to get a better feeling for what they are, what motivated their creation, how they organized communities, and what impact they had.

TIMIT

Automatic speech recognition is a machine learning problem of significant commercial interest. Its roots date back to the early twentieth century.[130]

Interestingly, speech recognition also features one of the oldest benchmarks datasets, the TIMIT (Texas Instruments/Massachusetts Institute for Technology) data. The creation of the dataset was funded through a 1986 DARPA program on speech recognition. In the mid-1980s, artificial intelligence was in the middle of a "funding winter," when many governmental and industrial agencies were hesitant to sponsor AI research because it often promised more than it could deliver. DARPA program manager Charles Wayne proposed a way around this problem by establishing more rigorous evaluation methods. Wayne enlisted the National Institute of Standards and Technology to create and curate shared datasets for speech, and he graded success in his program based on performance on recognition tasks on these datasets.

Many now credit Wayne's program with kick-starting a revolution of progress in speech recognition.[12,13,14] According to Church,

> It enabled funding to start because the project was glamour-and-deceit-proof, and to continue because funders could measure progress over time. Wayne's idea makes it easy to produce plots which help sell the research program to potential sponsors. A less obvious benefit of Wayne's idea is that it enabled hill climbing. Researchers who had initially objected to being tested twice a year began to evaluate themselves every hour.[13]

A first prototype of the TIMIT dataset was released in December 1988 on a CD-ROM. An improved release followed in October 1990. TIMIT already featured the training/test split typical for modern machine learning benchmarks. There's a fair bit we know about the creation of the data due to its thorough documentation.[131]

TIMIT features a total of about five hours of speech, composed of 6,300 utterances, specifically, 10 sentences spoken by each of 630 speakers. The sentences were drawn from a corpus of 2,342 sentences such as the following.

```
She had your dark suit in greasy wash water all year. (sa1)
Don't ask me to carry an oily rag like that. (sa2)
```

```
This was easy for us. (sx3)
Jane may earn more money by working hard. (sx4)
She is thinner than I am. (sx5)
Bright sunshine shimmers on the ocean. (sx6)
Nothing is as offensive as innocence. (sx7)
```

The TIMIT documentation distinguishes between eight major dialect regions in the United States:

> New England, Northern, North Midland, South Midland, Southern, New York City, Western, Army Brat (moved around)

Of the speakers, 70% are male and 30% are female. All native speakers of American English, the subjects were primarily employees of Texas Instruments at the time. Many of them were new to the Dallas area where they worked.

Racial information was supplied with the distribution of the data and coded as "White," "Black," "American Indian," "Spanish-American," "Oriental," and "Unknown." Of the 630 speakers, 578 were identified as White, 26 as Black, 2 as American Indian, 2 as Spanish-American, 3 as Oriental, and 17 as unknown.

Table 8.1: Demographic information about TIMIT speakers

	Male	Female	Total (%)
White	402	176	578 (91.7%)
Black	15	11	26 (4.1%)
American Indian	2	0	2 (0.3%)
Spanish-American	2	0	2 (0.3%)
Oriental	3	0	3 (0.5%)
Unknown	12	5	17 (2.6%)

The documentation notes:

> In addition to these 630 speakers, a small number of speakers with foreign accents or other extreme speech and/or hearing abnormalities were recorded as "auxiliary" subjects, but they are not included on the CD-ROM.

It comes as no surprise that early speech recognition models had significant demographic and racial biases in their performance.

Today, several major companies, including Amazon, Apple, Google, and Microsoft, all use speech recognition models in a variety of products from cell phone apps to voice assistants. Today, speech recognition lacks a major open benchmark that would support the training models competitive with the industrial counterparts. Industrial speech recognition pipelines are often complex systems that use proprietary data sources not a lot is known about. Nevertheless, even today's speech recognition systems continue to have racial biases.[132]

UCI Machine Learning Repository

The UCI Machine Learning Repository currently hosts more than 500 datasets, mostly for different classification and regression tasks. Most datasets are relatively small, many of them structured tabular datasets with few attributes.

The UCI Machine Learning Repository contributed to the adoption of the train-test paradigm in machine learning in the late 1980s. Langley recalls:

> The experimental movement was aided by another development. David Aha, then a PhD student at UCI, began to collect datasets for use in empirical studies of machine learning. This grew into the UCI Machine Learning Repository (`http://archive.ics.uci.edu/ml/`), which he made available to the community by FTP in 1987. This was rapidly adopted by many researchers because it was easy to use and because it let them compare their results to previous findings on the same tasks.[11]

Aha's PhD work involved evaluating nearest-neighbor methods, and he wanted to be able to compare the utility of his algorithms to decision tree induction algorithms, popularized by Ross Quinlan. Aha describes his motivation for building the UCI repository as follows.

> I was determined to create and share it, both because I wanted to use the datasets for my own research and because I thought it was ridiculous that the community hadn't fielded what should have been a useful service. I chose to use the simple attribute-value representation that Ross Quinlan was using so successfully for distribution with his TDIDT implementations.[133]

The UCI dataset was wildly successful, and partially responsible for the renewed interest in pattern recognition methods in machine learning. However, this success came with some detractors. By the mid 1990s, many were worried that evaluation-by-benchmark encouraged chasing state-of-the-art results and writing incremental papers. Aha reflects:

> By ICML-95, the problems "caused" by the repository had become popularly espoused. For example, at that conference Lorenza Saitta had, in an invited workshop that I co-organized, passionately decried how it allowed researchers to publish dull papers that proposed small variations of existing supervised learning algorithms

> and reported their small-but-significant incremental performance improvements in comparison studies.

Nonetheless, the UCI repository remains one of the most popular source for benchmark datasets in machine learning, and many of the early datasets still are used for benchmarking in machine learning research. The most popular dataset in the UCI repository is Ronald A. Fisher's Iris Data Set that Fisher collected for his 1936 paper on "The use of multiple measurements in taxonomic problems."

As of this writing, the second most popular dataset in the UCI repository is the *Adult* dataset. Extracted from the 1994 Census database, the dataset features nearly 50,000 instances describing individual in the United States, each having 14 attributes. The goal is to classify whether an individual earns more than 50,000 US dollars or less.

The Adult dataset became popular in the algorithmic fairness community, largely because it is one of the few publicly available datasets that features demographic information including *gender* (coded in binary as male/female) as well as *race* (coded as Amer-Indian-Eskimo, Asian-Pac-Islander, Black, Other, and White).

Unfortunately, the data has some idiosyncrasies that make it less than ideal for understanding biases in machine learning models.[134] Due to the age of the data, and the income cutoff at $50,000, almost all instances labeled *Black* are below the cutoff, as are almost all instances labeled *female*. Indeed, a standard logistic regression model trained on the data achieves about 85% accuracy overall, while the same model achieves 91% accuracy on Black instances, and nearly 93% accuracy on female instances. Likewise, the ROC curves for the latter two groups enclose actually more area than the ROC curve for male instances. This is a rather untypical situation since often machine learning models perform more poorly on historically disadvantaged groups.

Highleyman's data

The first machine learning benchmark dates back to the late 1950s. Few used it and even fewer still remembered it by the time benchmarks became widely used in machine learning in the late 1980s.

In 1959 at Bell Labs, Bill Highleyman and Louis Kamenstky designed a scanner to evaluate character recognition techniques.[135] Their goal was "to facilitate a systematic study of character-recognition techniques and an evaluation of methods prior to actual machine development." It was not clear at the time which part of the computations should be done in special purpose hardware and which parts should be done with more general computers. Highleyman later patented an optical character recognition (OCR) scheme that we recognize today as a convolutional neural network with convolutions optically computed as part of the scanning.[136]

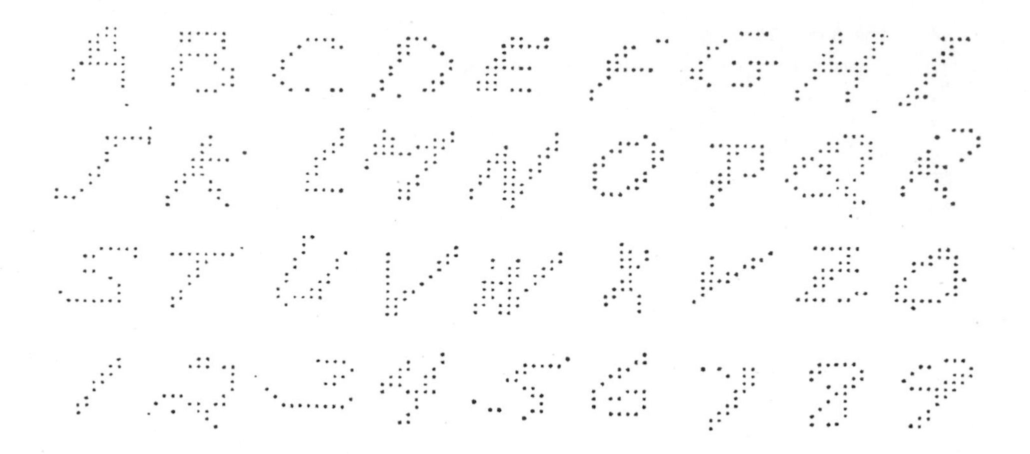

Figure 8.1: A look at Highleyman's data

Highleyman and Kamentsky used their scanner to create a data set of 1,800 alphanumeric characters. They gathered the 26 capital letters of the English alphabet and 10 digits from 50 different writers. Each character in their corpus was scanned in binary at a resolution of 12×12 and stored on punch cards that were compatible with the IBM 704, the first mass-produced computer with floating-point arithmetic hardware.

With the data in hand, Highleyman and Kamenstky began studying various proposed techniques for recognition. In particular, they analyzed a method of Woody Bledsoe's and published an analysis claiming to be unable to reproduce the results.[137] Bledsoe found the numbers to be considerably lower than he expected, and asked Highleyman to send him the data. Highleyman obliged, mailing the package of punch cards across the country to Sandia Labs. Upon receiving the data, Bledsoe conducted a new experiment. In what may be the first application of the train-test split, he divided the characters up, using 40 writers for training and 10 for testing. By tuning the hyperparameters, Bledsoe was able to achieve approximately 60% error.[138] Bledsoe also suggested that the high error rates were to be expected as Highleyman's data was too small. Prophetically, he declared that 1,000 alphabets might be needed for good performance.

By this point, Highleyman had also shared his data with Chao Kong Chow at the Burroughs Corporation (a precursor to Unisys). A pioneer of using decision theory for pattern recognition,[30] Chow built a pattern recognition system for characters. Using the same train-test split as Bledsoe, Chow obtained an error rate of 41.7%.[139]

Highleyman made at least six additional copies of the data he had sent to Bledsoe and Chow, and many researchers remained interested. He thus decided to publicly offer to send a copy to anyone willing to pay for the duplication and shipping fees.[140] Of course, the dataset was sent by US Postal Service. Electronic transfer didn't exist at the time, resulting in sluggish data transfer rates on the order of a few bits per minute.

Highleyman not only created the first machine learning benchmark. He authored the the first formal study of train-test splits[141] and proposed empirical risk minimization for pattern classification[31] as part of his 1961 dissertation. By 1963, however, Highleyman had left his research position at Bell Labs and abandoned pattern recognition research.

We don't know how many people requested Highleyman's data, but the total number of copies may have been less than twenty. Based on citation surveys, we determined there were at least six additional copies made after Highleyman's public offer for duplication, sent to researchers at UW Madison, CMU, Honeywell, SUNY Stony Brook, Imperial College in London, and Stanford Research Institute (SRI).

The SRI team of John Munson, Richard Duda, and Peter Hart performed some of the most extensive experiments with Highleyman's data.[142] A 1-nearest-neighbors baseline achieved an error rate of 47.5%. With a more sophisticated approach, they were able to do significantly better. They used a multi-class, piecewise linear model, trained using Kesler's multi-class version of the perceptron algorithm. Their feature vectors were 84 simple pooled edge detectors in different regions of the image at different orientations. With these features, they were able to get a test error of 31.7%, 10 points better than Chow. When restricted only to digits, this method recorded 12% error. The authors concluded that they needed more data, and that the error rates were "still far too high to be practical." They concluded that "larger and higher-quality datasets are needed for work aimed at achieving useful results." They suggested that such datasets "may contain hundreds, or even thousands, of samples in each class."

Munson, Duda, and Hart also performed informal experiments with humans to gauge the readability of Highleyman's characters. On the full set of alphanumeric characters, they found an average error rate of 15.7%, about 2x better than their pattern recognition machine. But this rate was still quite high and suggested the data needed to be of higher quality. They concluded that "an array size of at least 20×20 is needed, with an optimum size of perhaps 30×30."

Decades passed until such a dataset, the MNIST digit recognition task, was created and made widely available.

MNIST

The MNIST dataset contains images of handwritten digits. Its most common version has 60,000 training images and 10,000 test images, each having 28x28 grayscale pixels.

Figure 8.2: A sample of MNIST digits

MNIST was created by researchers Burges, Cortes, and LeCun from data by the National Institute of Standards and Technology (NIST). The dataset was introduced in a research paper in 1998 to showcase the use of gradient-based deep learning methods for document recognition tasks.[143] However, the authors released the dataset to provide a convenient benchmark of image data, in contrast to UCI's predominantly tabular data. The MNIST website states

> It is a good database for people who want to try learning techniques and pattern recognition methods on real-world data while spending minimal efforts on preprocessing and formatting.[144]

MNIST became a highly influential benchmark in the machine learning community. Two decades and over 30,000 citations later, researchers continue to use the data actively.

The original NIST data had the property that training and test data came from two different populations. The former featured the handwriting of 2,000 American Census Bureau employees, whereas the latter came from 500 American high school students.[145] The creators of MNIST reshuffled these two data sources and split them into training and test sets. Moreover, they scaled and centered the digits. The exact procedure to derive MNIST from NIST was lost, but recently reconstructed by matching images from both data sources.[146]

The original MNIST test set was of the same size as the training set, but the smaller test set became standard in research use. The 50,000 digits in the original test set that didn't make it into the smaller test set were later identified and dubbed *the lost digits*.[146]

From the beginning, MNIST was intended to be a benchmark used to compare the strengths of different methods. For several years, LeCun maintained an informal leaderboard on a personal website that listed the best accuracy numbers that different learning algorithms achieved on MNIST.

Table 8.2: A snapshot of the original MNIST leaderboard from February 2, 1999. Source: Internet Archive (Retrieved: December 4, 2020)

Method	Test error (%)
linear classifier (1-layer NN)	12.0
linear classifier (1-layer NN) [deskewing]	8.4
pairwise linear classifier	7.6
K-nearest-neighbors, Euclidean	5.0
K-nearest-neighbors, Euclidean, deskewed	2.4
40 PCA + quadratic classifier	3.3
1000 RBF + linear classifier	3.6
K-NN, Tangent Distance, 16x16	1.1
SVM deg 4 polynomial	1.1
Reduced Set SVM deg 5 polynomial	1.0
Virtual SVM deg 9 poly [distortions]	0.8
2-layer NN, 300 hidden units	4.7
2-layer NN, 300 HU, [distortions]	3.6
2-layer NN, 300 HU, [deskewing]	1.6
2-layer NN, 1000 hidden units	4.5
2-layer NN, 1000 HU, [distortions]	3.8
3-layer NN, 300+100 hidden units	3.05
3-layer NN, 300+100 HU [distortions]	2.5
3-layer NN, 500+150 hidden units	2.95
3-layer NN, 500+150 HU [distortions]	2.45
LeNet-1 [with 16x16 input]	1.7
LeNet-4	1.1
LeNet-4 with K-NN instead of last layer	1.1
LeNet-4 with local learning instead of ll	1.1
LeNet-5, [no distortions]	0.95
LeNet-5, [huge distortions]	0.85
LeNet-5, [distortions]	0.8
Boosted LeNet-4, [distortions]	0.7

In its capacity as a benchmark, it became a showcase for the emerging kernel methods of the early 2000s that temporarily achieved top performance on MNIST.[74] Today, it is not difficult to achieve less than 0.5% classification error with a wide range of convolutional neural network architectures. The best models classify all but a few pathological test instances correctly. As a result, MNIST is widely considered *too easy* for today's research tasks.

MNIST wasn't the first dataset of handwritten digits in use for machine learning research. Earlier, the US Postal Service (USPS) had released a dataset of 9,298 images (7,291 for training, and 2,007 for testing). The USPS data was actually a fair bit harder to classify than MNIST. A non-negligible fraction of the USPS digits look unrecognizable to humans,[147] whereas humans recognize essentially all digits in MNIST.

ImageNet

ImageNet is a large repository of labeled images that has been highly influential in computer vision research over the last decade. The image labels correspond to nouns from the WordNet lexical database of the English language. WordNet groups nouns into cognitive synonyms, called *synsets*. The words *car* and *automobile*, for example, would fall into the same synset. On top of these categories, WordNet provides a hierarchical structure according to a super-subordinate relationship between synsets. The synset for *chair*, for example, is a child of the synset for *furniture* in the WordNet hierarchy. WordNet existed before ImageNet and in part inspired the creation of ImageNet.

The initial release of ImageNet included about 5,000 image categories each corresponding to a synset in WordNet. These ImageNet categories averaged about 600 images per category.[148] ImageNet grew over time and its Fall 2011 release had reached about 32,000 categories.

The construction of ImageNet required two essential steps:

1. The first was the retrieval of candidate images for each synset.
2. The second step in the creation process was the labeling of images.

Scale was an important consideration due to the target size of the image repository.

This first step utilized available image databases with a search interface, specifically, Flickr. Candidate images were taken from the image search results associated with the synset nouns for each category.

For the second labeling step, the creators of ImageNet turned to Amazon's Mechanical Turk platform (MTurk). MTurk is an online labor market that allows individuals and corporations to hire on-demand workers to perform simple tasks. In this case, MTurk workers were presented with candidate images and had to decide whether or not the candidate image was indeed an image corresponding to the category that it was putatively associated with.

It is important to distinguish between this ImageNet database and a popular machine learning benchmark and competition, called ImageNet Large Scale Visual Recognition Challenge (ILSVRC), that was derived from it.[149] The competition was organized yearly from 2010 until 2017 to "measure the progress of computer vision for large scale image indexing for retrieval and annotation." In 2012, ILSVRC reached significant notoriety in both industry and academia when a deep neural network trained by Krizhevsky, Sutskever, and Hinton outperformed all other models by a significant margin.[124] This result—yet again an evaluation in a train-test paradigm—helped usher in the latest era of exuberant interest in machine learning and neural network models under the rebranding as *deep learning*.[150]

When machine learning practitioners say "ImageNet" they typically refer to the data used for the image classification task in the 2012 ILSVRC benchmark. The competition included other tasks, such as object recognition, but image classification has become the most popular task for the dataset. Expressions such as "a model trained on ImageNet" typically refer to an image classification model trained on the benchmark dataset from 2012.

Another common practice involving the ILSVRC data is *pre-training*. Often a practitioner has a specific classification problem in mind whose label set differs from the 1,000 classes present in the data. It's possible nonetheless to use the data to create useful features that can then be used in the target classification problem. Where ILSVRC enters real-world applications, it's often to support pre-training.

This colloquial use of the word ImageNet can lead to some confusion, not least because the ILSVRC-2012 dataset differs significantly from the broader database. It only includes a subset of 1,000 categories. Moreover, these categories are a rather skewed subset of the broader ImageNet hierarchy. For example, of these 1,000 categories only three are in the *person* branch of the WordNet hierarchy, specifically, *groom*, *baseball player*, and *scuba diver*. Yet, more than 100 of the 1,000 categories correspond to different dog breeds. The number is 118, to be exact, not counting wolves, foxes, and wild dogs that are also present among the 1,000 categories.

What motivated the exact choice of these 1,000 categories is not entirely clear. The apparent canine inclination, however, isn't just a quirk either. At the time, there was an interest in the computer vision community in making progress on prediction with many classes, some of which are very similar. This reflects a broader pattern in the machine learning community. The creation of datasets is often driven by an intuitive sense of what the technical challenges are for the field. In the case of ImageNet, *scale*, both in terms of the number of data points as well as the number of classes, was an important motivation.

The large scale annotation and labeling of datasets, such as we saw in the case of ImageNet, fall into a category of labor that anthropologist Gray and computer scientist Suri coined *Ghost Work* in their book of the same name.[151]

They point out:

> MTurk workers are the AI revolution's unsung heroes.

Indeed, ImageNet was labeled by about 49,000 MTurk workers from 167 countries over the course of multiple years.

Longevity of benchmarks

The holdout method is central to the scientific and industrial activities of the machine learning community. Thousands of research papers have been written that report numbers on popular benchmark data, such as MNIST, CIFAR-10, and ImageNet. Often extensive tuning and hyperparameter optimization went into each such research project to arrive at the final accuracy numbers reported in the paper.

Does this extensive reuse of test sets not amount to what Duda and Hart call the "training on the testing data" problem? If so, how much of the progress that has been made is real, and how much amounts of overfitting to the test data?

To answer these questions we will develop some more theory that will help us interpret the outcome of empirical meta studies into the longevity of machine learning benchmarks.

The problem of adaptivity

Model building is an iterative process where the performance of a model informs subsequent design choices. This iterative process creates a closed feedback loop between the practitioner and the test set. In particular, the models the practitioner chooses are not independent of the test set, but, rather, *adaptive*.

Adaptivity can be interpreted as a form of overparameterization. In an adaptively chosen sequence of predictors $f_1, \ldots, f_k$, the kth predictor has the ability to incorporate at least $k-1$ bits of information about the performance of previously chosen predictors. This suggests that as $k \geq n$, the statistical guarantees of the holdout method become vacuous. This intuition is formally correct, as we will see.

To reason about adaptivity, it is helpful to frame the problem as an interaction between two parties. One party holds the dataset S. Think of this party as implementing the holdout method. The other party, which we call *analyst*, can *query* the dataset by requesting the empirical risk $R_S[f]$ of a given predictor f on the dataset S. The parties interact for some number k of rounds, thus creating a sequence of adaptively chosen predictors $f_1, \ldots, f_k$. Keep in mind that this sequence depends on the dataset! In particular, when S is drawn

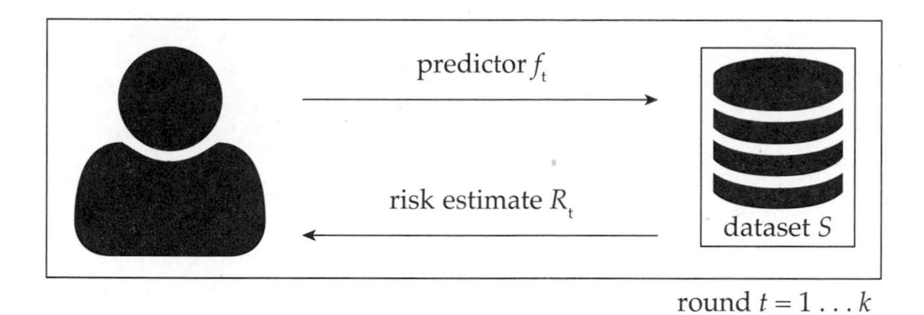

Figure 8.3: The adaptive analyst model

at random, $f_2, \ldots, f_k$ become random variables, too, that are in general not independent of each other.

In general, the estimate R_t returned by the holdout mechanism at step t need not be equal to $R_S[f_t]$. We can often do better than the standard holdout mechanism by limiting the information revealed by each response. Throughout this chapter, we restrict our attention to the case of the zero-one loss and binary prediction, although the theory extends to other settings.

As it turns out, the guarantee of the standard holdout mechanism in the adaptive case is exponentially worse in k compared with the non-adaptive case. Indeed, there is a fairly natural sequence of k adaptively chosen predictors, resembling the practice of ensembling, on which the empirical risk is off by at least $\Omega(\sqrt{k/n})$. This is a lower bound on the gap between risk and empirical risk in the adaptive setting. Contrast this with the $O(\sqrt{\log(k)/n})$ upper bound that we observed for the standard holdout mechanism in the non-adaptive case. We present the idea for the zero-one loss in a binary prediction problem.

Overfitting by ensembling:

1. Choose k of random binary predictors $f_1, \ldots, f_k$.
2. Compute the set $I = \{i \in [k] : R_S[f_i] < 1/2\}$.
3. Output the predictor $f = \text{majority}\{f_i : i \in I\}$ that takes a majority vote over all the predictors computed in the second step.

The key idea of the algorithm is to select all the predictors that have accuracy strictly better than random guessing. This selection step creates a bias that gives each selected predictor an advantage over random guessing. The majority vote in the third step amplifies this initial advantage into a larger advantage that grows with k. The next proposition confirms that indeed this strategy finds a predictor whose empirical risk is bounded away from $1/2$ (random guessing) by a margin of $\Omega(\sqrt{k/n})$. Since the predictor does nothing but take a majority vote over random functions, its risk is of course no better than $1/2$.

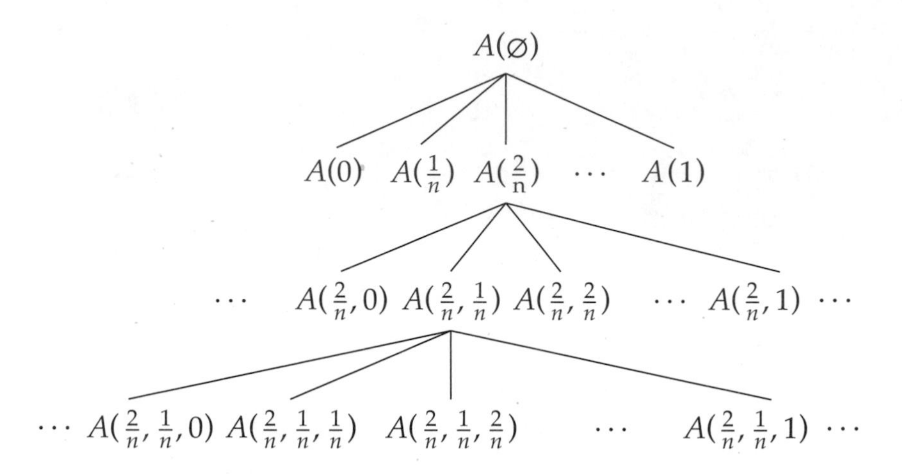

Figure 8.4: An adaptive analyst as a tree of depth k and degree $n+1$

Proposition 9. *For sufficiently large $k \leq n$, overfitting by ensembling returns a predictor f whose classification error satisfies, with probability $1/3$,*

$$R_S[f] \leq \frac{1}{2} - \Omega(\sqrt{k/n}).$$

In particular, $\Delta_{\text{gen}}(f) \geq \Omega(\sqrt{k/n})$.

We also have a nearly matching upper bound that essentially follows from a Hoeffding's concentration inequality just like the cardinality bound in the previous chapter. However, in order to apply Hoeffding's inequality we first need to understand a useful idea about how we can analyze the adaptive setting.

The idea is that we can think of the interaction between a fixed analyst $\mathcal{A}$ and the dataset as a *tree*. Each node corresponds to the predictor the analyst chooses based on the responses seen so far. The root node is labeled by $f_1 = \mathcal{A}(\varnothing)$, i.e., the first function that the analyst queries without any input. The response $R_S[f_1]$ takes on $n+1$ possible values. This is because we consider the zero-one loss, which can only take the values $\{0, 1/n, 2/n, \ldots, 1\}$. Each possible response value a_1 creates a new child node in the tree corresponding to the function $f_2 = \mathcal{A}(a_1)$ that the analyst queries upon receiving answer a_1 to the first query f_1. We recursively continue the process until we build up a tree of depth k and degree $n+1$ at each node.

Note that this tree only depends on the analyst and how it responds to possible query answers; it does not depend on the actual query answers we get out of the sample S. The tree is therefore data-independent. This argument is useful in proving the following proposition.

Proposition 10. *For any sequence of k adaptively chosen predictors $f_1, \ldots, f_k$, the holdout method satisfies, with probability* $2/3$,

$$\max_{1 \leq t \leq k} \Delta_{\text{gen}}(f_t) \leq O\left(\sqrt{k \log(n+1)/n}\right).$$

Proof. The adaptive analyst defines a tree of depth k and degree $n+1$. Let F be the set of functions appearing at any of the nodes in the tree. Note that $|F| \leq (n+1)^k$.

Since this set of functions is data-independent, we can apply the cardinality bound from the previous chapter to argue that the maximum generalization gap for any function in F is bounded by $O(\sqrt{\log|F|/n})$ with any constant probability. But the functions $f_1, \ldots, f_k$ are contained in F by construction. Hence, the claim follows.

□

These propositions show that the principal concern of "training on the testing data" is not unfounded. In the worst case, holdout data can lose its guarantees rather quickly. If this pessimistic bound manifested in practice, popular benchmark datasets would quickly become useless. But does it?

Replication efforts

In recent replication efforts, researchers carefully recreated new test sets for the CIFAR-10 and ImageNet classification benchmarks, created according to the very same procedure as the original test sets. The researchers then took a large collection of representative models proposed over the years and evaluated all of them on the new test sets. In the case of MNIST, researchers used the *lost digits* as a new test set, since these digits hadn't been used in almost all of the research on MNIST.

The results of these studies teach us a couple of important lessons that we will discuss in turn.

First, all models suffer a significant drop in performance on the new test set. The accuracy on the new data is substantially lower than on the old data. This shows that these models *transfer* surprisingly poorly from one dataset to a very similar dataset that was constructed in much the same way as the original data. This observation resonates with a robust fact about machine learning. Model fitting will do exactly that. The model will be good on exactly the data it is trained on, but there is no good reason to believe that it will perform well on other data. Generalization as we cast it in the preceding chapter is thus about *interpolation*. It's about doing well on more data from the same source. It is decidedly *not* about doing well on data from other sources.

The second observation is relevant to the question of adaptivity; it's a bit more subtle. The scatter plots admit a clean linear fit with positive slope. In

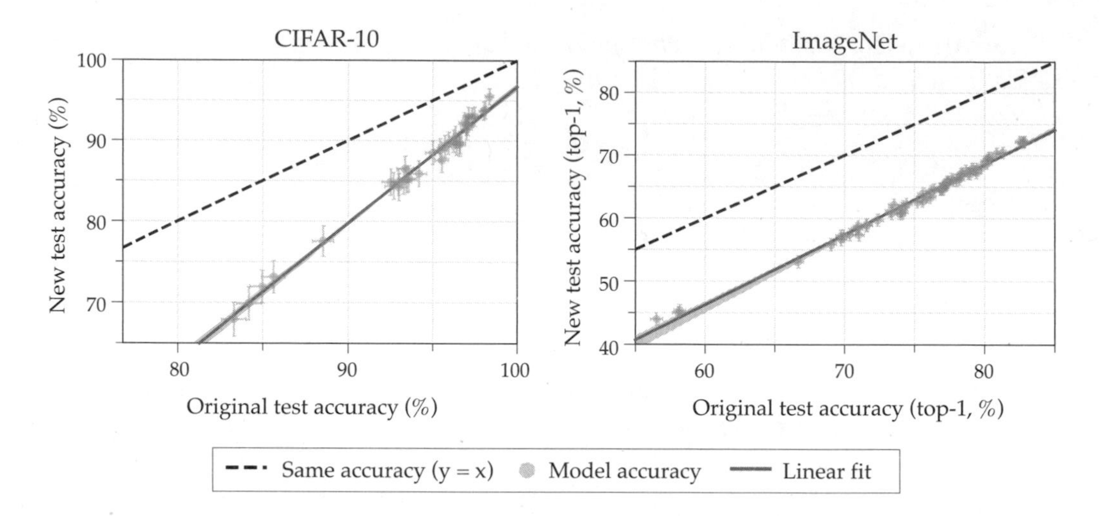

Figure 8.5: Model accuracy on the original test sets vs. new test sets for CIFAR-10 and ImageNet. Each data point corresponds to one model in a test bed of representative models.

other words, the better a model is on the old test set, the better it is on the new test set, as well. But notice that newer models, i.e., those with higher performance on the original test set, had *more* time to adapt to the test set and to incorporate more information about it. Nonetheless, the better a model performed on the old test set, the better it performs on the new set. Moreover, on CIFAR-10 we even see clearly that the absolute performance drop diminishes with increasing accuracy on the old test set. In particular, if our goal was to do well on the new test set, seemingly our best strategy is to continue to inch forward on the old test set. This might seem counterintuitive.

We will discuss each of these two observations in more detail, starting with the one about adaptivity.

Benign adaptivity

The experiments we just discussed suggest that the effect of adaptivity was more benign than our previous analysis suggested. This raises the question of what it is that prevents more serious overfitting. There are a number of pieces to the puzzle that researchers have found. Here we highlight two.

The main idea behind both mechanisms that damp adaptivity is that the set of possible nodes in the adaptive tree may be much less than n^k because of empirical conventions. The first mechanism is *model similarity*. Effectively, model similarity notes that the leaves of the adaptive tree may be producing similar predictions, and hence the adaptivity penalty is smaller than our worst-case count. The second mechanism is the *leaderboard principle*. This more subtle principle states that since publication biases force researchers to chase state-of-the-art results, they only publish models if they see significant improvements over prior models.

While we don't believe that these two mechanisms explain the entirety of why overfitting is not observed in practice, they significantly reduce the effects of adaptivity. As we said, these are two examples of norms in machine learning practice that diminish the effects of overfitting.

Model similarity

Naively, there are 2^n total assignments of binary labels to a dataset of size n. But how many such labeling assignments do we see in practice? We do not solve pattern recognition problems using the ensembling attack described above. Rather, we use a relatively small set of function approximation architectures, and tune the parameters of these architectures. While we have seen that these architectures can yield any of the 2^n labeling patterns, we expect that a much smaller set of predictions is returned in practice when we run standard empirical risk minimization.

Model similarity formalizes this notion as follows. Given an adaptively chosen sequence of predictors $f_1, \ldots, f_k$, we have a corresponding sequence of empirical risks $R_1, \ldots, R_k$.

Definition 11. *We say that a sequence of models $f_1, \ldots, f_k$ are ζ-*similar *if for all pairs of models f_i and f_j with empirical risks $R_i \leq R_j$, respectively, we have*

$$\mathbb{P}\left[\{f_j(x) = y\} \cap \{f_i(x) \neq y\}\right] \leq \zeta \,.$$

This definition states that there is low probability of a model with small empirical risk misclassifying an example where a model with higher empirical risk was correct. It effectively grades the set of *examples* as being easier or harder to classify, and suggests that models with low risk usually get the easy examples correct.

Though simple, this definition is sufficient to reduce the size of the adaptive tree, thus leading to a better theoretical bound.[152] Empirically, deep learning models appear to have high similarity beyond what follows from their accuracies.[152]

The definition of similarity can also help explain the scatter plots we saw previously: When we consider the empirical risks of high similarity models on two different test sets, the scatter plot of the (R_i, R_i') pairs cluster around a line.[153]

The leaderboard principle

The leaderboard principle postulates that *a researcher only cares if their model improved over the previous best or not.* This motivates a notion of *leaderboard error* where the holdout method is only required to track the risk of the best performing model over time, rather than the risk of all models ever evaluated.

Definition 12. *Given an adaptively chosen sequence of predictors $f_1, \dots, f_k$, we define the* leaderboard error *of a sequence of estimates $R_1, \dots, R_k$ as*

$$\operatorname{lberr}(R_1, \dots, R_k) = \max_{1 \le t \le k} \left| \min_{1 \le i \le t} R[f_i] - R_t \right| .$$

We discuss an algorithm called the Ladder algorithm that achieves small leaderboard error. The algorithm is simple. For each given predictor, it compares the empirical risk estimate of the predictor to the previously smallest empirical risk achieved by any predictor encountered so far. If the loss is below the previous best by some margin, it announces the empirical risk of the current predictor and notes it as the best seen so far. Importantly, if the loss is not smaller by a margin, the algorithm releases no new information and simply continues to report the previous best.

Again, we focus on risk with respect to the zero-one loss, although the ideas apply more generally.

Input: Dataset S, threshold $\eta > 0$

- Assign initial leaderboard error $R_0 := 1$.
- For each round $t = 1, 2 \dots$:

  1. Receive predictor $f_t \colon X \to Y$.
  2. If $R_S[f_t] < R_{t-1} - \eta$, update leaderboard error to $R_t := R_S[f_t]$. Else keep previous leaderboard error $R_t := R_{t-1}$.
  3. Output the leaderboard error R_t.

The next theorem follows from a variant of the adaptive tree argument we saw earlier, in which we carefully prune the tree and bound its size.

Theorem 9. *For a suitably chosen threshold parameter, for any sequence of adaptively chosen predictors $f_1, \dots, f_k$, the Ladder algorithm achieves with probability $1 - o(1)$:*

$$\operatorname{lberr}(R_1, \dots, R_k) \le O\left(\frac{\log^{1/3}(kn)}{n^{1/3}} \right) .$$

Proof. Set $\eta = \log^{1/3}(kn)/n^{1/3}$. With this setting of η, it suffices to show that with probability $1 - o(1)$ we have for all $t \in [k]$ the bound $|R_S[f_t] - R[f_t]| \le O(\eta) = O(\log^{1/3}(kn)/n^{1/3})$.

Let $\mathcal{A}$ be the adaptive analyst generating the function sequence. The algorithm $\mathcal{A}$ naturally defines a rooted tree $\mathcal{T}$ of depth k recursively defined as follows:

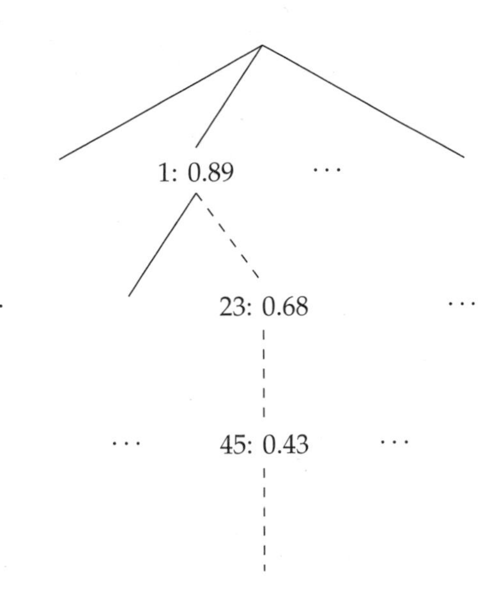

Figure 8.6: Low bit encoding of the adaptive tree. Dashed lines correspond to rounds with no update.

1. The root is labeled by $f_1 = \mathcal{A}(\emptyset)$.
2. Each node at depth $1 < i \leq k$ corresponds to one realization $(h_1, r_1, \ldots, h_{i-1}, r_{i-1})$ of the tuple of random variables $(f_1, R_1, \ldots, f_{i-1}, R_{i-1})$ and is labeled by $h_i = \mathcal{A}(h_1, r_1, \ldots, h_{i-1}, r_{i-1})$. Its children are defined by each possible value of the output R_i of Ladder Mechanism on the sequence $h_1, r_1, \ldots, r_{i-1}, h_i$.

Let $B = (1/\eta + 1)\log(4k(n+1))$. We claim that the size of the tree satisfies $|\mathcal{T}| \leq 2^B$. To prove the claim, we will uniquely encode each node in the tree using B bits of information. The claim then follows directly.

The compression argument is as follows. We use $\lceil \log k \rceil \leq \log(2k)$ bits to specify the depth of the node in the tree. We then specify the index of each $i \in [k]$ for which the Ladder algorithm performs an "update" so that $R_i \leq R_{i-1} - \eta$ together with the value R_i. Note that since $R_i \in [0,1]$ there can be at most $\lceil 1/\eta \rceil \leq (1/\eta) + 1$ many such steps. This is because the loss is lower bounded by 0 and decreases by η each time there is an update.

Moreover, there are at most $n+1$ many possible values for R_i, since we're talking about the zero-one loss on a dataset of size n. Hence, specifying all such indices requires at most $(1/\eta + 1)(\log(n+1) + \log(2k))$ bits. These bits of information uniquely identify each node in the graph, since for every index i not explicitly listed we know that $R_i = R_{i-1}$. The total number of bits we used is:

$$(1/\eta + 1)(\log(n+1) + \log(2k)) + \log(2k) \leq (1/\eta + 1)\log(4k(n+1)) = B\,.$$

This establishes the claim we made. The theorem now follows by applying a union bound over all nodes in $\mathcal{T}$ and using Hoeffding's inequality for each fixed node. Let $\mathcal{F}$ be the set of all functions appearing in $\mathcal{T}$. By a union bound, we have

$$\begin{aligned}\mathbb{P}\{\exists f \in \mathcal{F}\colon\ |R[f] - R_S[f]| > \epsilon\} &\leq 2|\mathcal{F}|\exp(-2\epsilon^2 n)\\ &\leq 2\exp(-2\epsilon^2 n + B)\,.\end{aligned}$$

Verify that by putting $\epsilon = 5\eta$, this expression can be upper bounded by $2\exp(-n^{1/3}) = o(1)$, and thus the claim follows.

□

Harms associated with data

The collection and use of data often raises serious ethical concerns. We will walk through some that are particularly relevant to machine learning.

Representational harm and biases

As we saw earlier, we have no reason to expect a machine learning model to perform well on any population that differs significantly from the training data. As a result, underrepresentation or misrepresentation of populations in the training data has direct consequences on the performance of any model trained on the data.

In a striking demonstration of this problem, Buolamwini and Gebru[154] point out that two facial analysis benchmarks, IJB-A and Adience, overwhelmingly featured lighter-skinned subjects. Introducing a new facial analysis dataset, which is balanced by gender and skin type, Buolamwini and Gebru demonstrated that commercial face recognition software misclassified darker-skinned females at the highest rate, while misclassifying lighter-skinned males at the lowest rate.

Images are not the only domain where this problem surfaces. Models trained on text corpora reflect the biases and stereotypical representations present in the training data. A well-known example is the case of word embeddings. Word embeddings map words in the English language to a vector representation. This representation can then be used as a feature representation for various other downstream tasks. A popular word embedding method is Google's `word2vec` tool that was trained on a corpus of Google news articles. Researchers demonstrated that the resulting word embeddings encoded stereotypical gender representations of the form *man is to computer programmer as woman is to homemaker*.[155] Findings like these motivated much work on *debiasing* techniques that aim to remove such biases from the learned representation.

However, there is doubt whether such methods can successfully remove bias after the fact.[156]

Privacy violations

The Netflix Prize was one of the most famous machine learning competitions. Starting on October 2, 2006, the competition ran for nearly three years, ending with a grand prize of $1 million, announced on September 18, 2009. Over the years, the competition saw 44,014 submissions from 5169 teams.

The Netflix training data contained roughly 100 million movie ratings from nearly 500,000 Netflix subscribers on a set of 17,770 movies. Each data point corresponds to a tuple <user, movie, date of rating, rating>. At about 650 megabytes in size, the dataset was just small enough to fit on a CD-ROM, but large enough to be pose a challenge at the time.

The Netflix data can be thought of as a matrix with $n = 480,189$ rows and $m = 17,770$ columns. Each row corresponds to a Netflix subscriber and each column to a movie. The only entries present in the matrix are those for which a given subscriber rated a given movie with rating in $\{1,2,3,4,5\}$. All other entries—that is, the vast majority—are *missing*. The objective of the participants was to predict the missing entries of the matrix, a problem known as *matrix completion*, or *collaborative filtering* somewhat more broadly. In fact, the Netflix challenge did so much to popularize this problem that it is sometimes called the *Netflix problem*. The idea is that if we could predict missing entries, we'd be able to recommend unseen movies to users accordingly.

The holdout data that Netflix kept secret consisted of about three million ratings. Half of them were used to compute a running leaderboard throughout the competition. The other half determined the final winner.

The Netflix competition was hugely influential. Not only did it attract significant participation, it also fueled much academic interest in collaborative filtering for years to come. Moreover, it popularized the competition format as an appealing way for companies to engage with the machine learning community. A startup called Kaggle, founded in April 2010, organized hundreds of machine learning competitions for various companies and organizations before its acquisition by Google in 2017.

But the Netflix competition became infamous for another reason. Although Netflix had replaced usernames by random numbers, researchers Narayanan and Shmatikov were able to reidentify many of the Netflix subscribers whose movie ratings were in the dataset.[157] In a nutshell, their idea was to link movie ratings in the Netflix dataset with publicly available movie ratings on IMDB, an online movie database. Some Netflix subscribers had also publicly rated an overlapping set of movies on IMDB under their real name. By matching movie ratings between the two sources of information, Narayanan and Shmatikov succeeded in associating anonymous users in the Netflix data with real names from IMDB. In the privacy literature, this is called a *linkage attack* and it's one of the many ways that seemingly anonymized data can be deanonymized.[158]

What followed were multiple class action lawsuits against Netflix, as well as an inquiry by the Federal Trade Commission over privacy concerns. As a consequence, Netflix canceled plans for a second competition, which it had announced on August 6, 2009.

To this day, privacy concerns are a highly legitimate obstacle to public data release and dataset creation. Deanonymization techniques are mature and efficient. There provably is no algorithm that can take a dataset and provide a rigorous privacy guarantee to all participants, while being useful for all analyses and machine learning purposes. Dwork and Roth call this the Fundamental Law of Information Recovery: *"overly accurate answers to too many questions will destroy privacy in a spectacular way."*[159]

Copyright

Privacy concerns are not the only obstruction to creating public datasets and using data for machine learning purposes. Almost all data sources are also subject to copyright. Copyright is a type of intellectual property, protected essentially worldwide through international treaties. It gives the creator of a piece of work the exclusive right to create copies of it. Copyright expires only decades after the creator dies. Texts, images, videos, digital or not, are all subject to copyright. Copyright infringement is a serious crime in many countries.

The question of how copyright affects machine learning practice is far from settled. Courts have yet to set precedents on whether copies of content that feed into machine learning training pipelines may be considered copyright infringement.

Legal scholar Levendowski argues that copyright law biases creators of machine learning systems toward "biased, low-friction data." These are data sources that carry a low risk of creating a liability under copyright law, but carry various biases in the data that affect how the resulting models perform.[160]

One source of low-friction data is what is known as "public domain." Under current US law, works enter public domain 75 years after the death of the copyright holder. This means that most public domain works were published prior to 1925. If the creator of a machine learning system relied primarily on public domain works for training, it would likely bias the data toward older content.

Another example of a low-friction dataset is the *Enron email corpus* that contains 1.6 million emails sent among Enron employees over the course of multiple years leading up to the collapse of the company in 2001. The corpus was released by the Federal Energy Regulatory Commission (FERC) in 2003, following its investigation into the serious accounting fraud case that became known as "Enron scandal." The Enron dataset is one of the few available large data sources of emails sent between real people. Even though the data were released by regulators to the public, that doesn't mean that they are "public domain." However, it is highly unlikely that a former Enron employee might sue for copyright infringement. The dataset has numerous biases. The emails are two decades old, and sent by predominantly male senior managers in a particular business sector.

An example of a dataset that is *not* low-friction is the corpus of news articles that became the basis for Google's famous word embedding tool called word2vec that we mentioned earlier. Due to copyright concerns with the news articles contained in the corpus, the dataset was never released, only the trained model.

Problem framing and comparisons with humans

A long-standing ambition of artificial intelligence research is to match or exceed human cognitive abilities by an algorithm. This desire often leads to comparisons between humans and machines on various tasks. Judgments about human accuracy often also enter the debate around when to use statistical models in high-stakes decision making settings.

The comparison between human decision makers and statistical models is by no means new. For decades, researchers have compared the accuracy of human judgments with that of statistical models.[161]

Even within machine learning, the debate dates way back. A 1991 paper by Bromley and Sackinger explicitly compared the performance of artificial neural networks to a measure of human accuracy on the USPS digits dataset that predates the famous MNIST data.[147] A first experiment put the human accuracy at 2.5%, a second experiment found the number 1.51%, while a third reported the number 2.37%.[162]

Comparison with so-called *human baselines* has since become widely accepted in the machine learning community. The Electronic Frontier Foundation (EFF), for example, hosts a major repository of AI progress measures that compares the performance of machine learning models to *reported human accuracies* on numerous benchmarks.[163]

For the ILSVRC 2012 data, the reported human error rate is 5.1%. To be precise, this number is referring to the fraction of times that the correct image label was not contained in the top five predicted labels. This often quoted number corresponds to the performance of a single human annotator who was "trained on 500 images and annotated 1,500 test images." A second annotator who was "trained on 100 images and then annotated 258 test images" achieved an error rate of 12%.[149]

Based on this number of 5.1%, researchers announced in 2015 that their model was "the first to surpass human-level performance."[164] Not surprisingly, this claim received significant attention throughout the media.

However, a later more careful investigation into "human accuracy" on ImageNet revealed a very different picture.[165] The researchers found that only models from 2020 are actually on par with the strongest human labeler. Moreover, when restricting the data to 590 object classes out of 1,000 classes in total, the best human labeler performed much better, at less than 1% error, than even the best predictive models. Recall, that the ILSVRC 2012 data featured 118 different dog breeds alone, some of which are extremely hard to distinguish for anyone who is not a trained dog expert. In fact, the researchers had to consult with experts from the American Kennel Club (AKC) to disambiguate challenging cases of different dog breeds. Simply removing dog classes alone increases the performance of the best human labeler to less than 1.3% error.

There is another troubling fact. Small variations in the data collection protocol turn out to have a significant effect on the performance of machine predictors: "the accuracy scores of even the best image classifiers are still highly sensitive to minutiae of the data cleaning process."[166]

These results cast doubt not only on how me measure "human accuracy," but also on the validity of the presumed theoretical construct of "human accuracy" itself. It is helpful to take a step back and reflect on measurement more broadly. Recall from Chapter 4 that the field of measurement theory distinguishes between a measurement procedure and the target *construct* that we wish to measure. For any measurement to be valid, the target construct has to be *valid* in the first place.

However, the machine learning community has adopted a rather casual approach to measuring human accuracy. Many researchers assume that the construct of *human accuracy* exists unambiguously and it is whatever number comes out of some ad hoc testing protocol for some set of human beings. These ad hoc protocols often result in anecdotal comparisons of questionable scientific value.

There is a broader issue with the idea of *human accuracy*. The notion presupposes that we have already accepted the prediction task to be the definitive task that we ought to solve, thus forgoing alternative solutions. But in many cases the problem formulation in itself is the subject of normative debate.

Consider the case of predicting *failure to appear in court*. This prediction problem is at the center of an ongoing criminal justice reform in the United States. Many proponents seek to replace, or at least augment, human judges by statistical models that predict whether or not a defendant would fail to appear in court if released ahead of a future trial. Defendants of high risk are jailed, often for months without a verdict, until their court appointment. An alternative to prediction is to understand the *causes* of failure to appear in court, and to specifically address these. We will turn to causality in subsequent chapters, where we will see that it often provides an important alternative to prediction.

As it turns out, defendants often fail to appear in court for lack of transportation, lack of childcare, inflexible work hours, or simply too many court appointments. Addressing these fundamental problems, in fact, is part of a settlement in Harris County, Texas.

To conclude, invalid judgments about human performance relative to machines are not just scientific errors, they also have the potential to create narratives that support poor policy choices in high-stakes policy questions around the use of predictive models in consequential decisions.

Toward better data practices

The practices of data collection and dataset creation in the machine learning community leave much room for improvement. We close this chapter highlighting a few practices that can be immediately adopted.

Data annotation

Many existing datasets in machine learning are poorly documented, and details about their creation are often missing. This leads to a range of issues from lack of reproducibility and concerns of scientific validity to misuse and ethical concerns. Fortunately, there is some emerging literature on how to better execute and document the creation of datasets for machine learning.

Datasheets for datasets is an initiative to promote a more detailed and systematic annotation for datasets.[167] A datasheet requires the creator of a dataset to answer questions relating to several areas of interest: Motivation, composition, collection process, preprocessing/cleaning/labeling, uses, distribution, maintenance.

One goal is that the process of creating a datasheet will help anticipate ethical issues with the dataset. But datasheets also aim to make data practices more reproducible, and help practitioners select more adequate data sources.

Going a step beyond datasheets, researchers Jo and Gebru[168] draw lessons from archival and library sciences for the construction and documentation of machine learning datasets. These lessons draw attention to issues of consent, inclusivity, power, transparency, ethics, and privacy.

Lessons from measurement

Measurement theory is an established science with ancient roots. In short, measurement is about assigning numbers to objects in the real world in a way that reflects relationships between these objects. Measurement draws an important distinction between a *construct* that we wish to measure and the measurement procedure that we used to create a numerical representation of the construct.

For example, we can think of a well-designed math exam as measuring the mathematical abilities of a student. A student with greater mathematical ability than another is expected to score higher on the exam. Viewed this way, an exam is a *measurement procedure* that assigns numbers to students. The *mathematical ability* of a student is the construct we hope to measure. We desire that the ordering of these numbers reflects the sorting of students by their mathematical abilities. A measurement procedure operationalizes a construct.

Recall that in a prediction problem we have covariates X from which we're trying to predict a variable Y. This variable Y is what we call the *target variable* in our prediction problem. The definition and choice of a target variable is one way that measurement theory becomes relevant to machine learning practice.

Consider a machine learning practitioner who attempts to classify the sentiment of a paragraph of text as "toxic" or not. In the language of measurement, "toxicity" is a construct. Whatever labeling procedure the practitioner comes up with can be thought of as a measurement procedure that implicitly or explicitly operationalizes this construct. Before resorting to ad hoc labeling or survey procedures, machine learning practitioners should survey available research.

A poor target variable cannot be ironed out with additional data. In fact, the more data we feed into our model, the better it gets at capturing the flawed target variable. Improved data quality or diversity are no cure either.

Formal fairness criteria that involve the target variable—separation and sufficiency are two prominent examples—are either meaningless or downright misleading when the target variable itself is the locus of discrimination. Recall from Chapter 2 that separation requires the protected attribute to be independent of the prediction conditional on the target variable. Sufficiency requires the target variable to be independent of the protected attribute given the prediction.

To get a better grasp on what makes a target variable more or less problematic, consider a few examples.

1. Predicting the value of the Standard and Poor's 500 Index (S&P 500) at the close of the New York Stock Exchange tomorrow.
2. Predicting whether an individual is going to default on a loan.
3. Predicting whether an individual is going to commit a crime.

The first example is rather innocuous. It references a fairly robust target variable, even though it relies on a number of social facts.

The second example is a common application of statistical modeling that underlies much of modern credit scoring in the United States. At first sight a default event seems like a clear-cut target variable. But the reality is different. In a public dataset released by the Federal Reserve,[169] default events are coded by a so-called *performance* variable that measures a *serious delinquency in at least one credit line of a certain time period*. More specifically, the report states that the

> measure is based on the performance of new or existing accounts and measures whether individuals have been late 90 days or more on one or more of their accounts or had a public record item or a new collection agency account during the performance period.[169]

Our third example runs into the most concerning measurement problem. How do we determine if an individual committed a crime? What we can determine with certainty is whether or not an individual was arrested and found guilty of a crime. But this depends crucially on who is likely to be policed in the first place and who is able to maneuver the criminal justice system successfully following an arrest.

Sorting out what a good target variable is, in full generality, can involve the whole apparatus of measurement theory. The scope of measurement theory, however, goes beyond defining reliable and valid target variables for prediction. Measurement comes in whenever we create features for a machine learning problem and should therefore be an essential part of the data creation process.

Judging the quality of a measurement procedure is a difficult task. Measurement theory has two important conceptual frameworks for arguing about what makes measurement good. One is *reliability*. The other is *validity*.

Reliability describes the differences observed in multiple measurements of the same object under identical conditions. Thinking of the measurement variable as a random variable, reliability is about the variance between independent identically distributed measurements. As such, reliability can be analogized with the statistical notion of variance.

Validity is concerned with how well the measurement procedure in principle captures the concept that we try to measure. If reliability is analogous to variance, it is tempting to see validity as analogous to *bias*. But the situation is a bit more complicated. There is no simple formal criterion that we could use to establish validity. In practice, validity is based to a large extent on human expertise and subjective judgments.

One approach to formalize validity is to ask how well a score predicts some external criterion. This is called *external validity*. For example, we could judge a measure of creditworthiness by how well it predicts default in a lending scenario. While external validity leads to concrete technical criteria, it essentially identifies good measurement with predictive accuracy. However, that's certainly not all there is to validity.

Construct validity is a framework for discussing validity that includes numerous different types of evidence. Messick highlights six aspects of construct validity:

- Content: How well does the content of the measurement instrument, such as the items on a questionnaire, measure the construct of interest?
- Substantive: Is the construct supported by a sound theoretical foundation?

- Structural: Does the score express relationships in the construct domain?
- Generalizability: Does the score generalize across different populations, settings, and tasks?
- External: Does the score successfully predict external criteria?
- Consequential: What are the potential risks of using the score with regard to bias, fairness, and distributive justice?

Of these different criteria, external validity is the one most familiar to the machine learning practitioner. But machine learning practice would do well to embrace the other, more qualitative, criteria as well. Ultimately, measurement forces us to grapple with the often surprisingly uncomfortable question: What are we even trying to do when we predict something?

Limits of data and prediction

Machine learning fails in many scenarios and it's important to understand the failure cases as much as the success stories.

The Fragile Families Challenge was a machine learning competition based on the Fragile Families and Child Wellbeing study (FFCWS).[170] Starting from a random sample of hospital births between 1998 and 2000, the FFCWS followed thousands of American families over the course of 15 years, collecting detailed information about the families' children, their parents, educational outcomes, and the larger social environment. Once a family agreed to participate in the study, data were collected when the child was born, and then at ages 1, 3, 5, 9, and 15.

The Fragile Families Challenge took concluded in 2017. The underlying dataset for the competition contains 4,242 rows, one for each family, and 12,943 columns, one for each variable plus an ID number of each family. Of the 12,942 variables, 2,358 are constant (i.e., had the same value for all rows), mostly due to redactions for privacy and ethics concerns. Of the approximately 55 million (4,242 x 12,942) entries in the dataset, about 73% do not have a value. Missing values have many possible reasons, including non-response of surveyed families, dropout of study participants, as well as logical relationships between features that imply certain fields are missing depending on how others are set. There are six outcome variables, measured at age 15: *1) child grade point average (GPA), 2) child grit, 3) household eviction, 4) household material hardship, 5) caregiver layoff, and 6) caregiver participation in job training.*

The goal of the competition was to predict the value of the outcome variables at age 15 given the data from ages 1 through 9. As is common for competitions, the challenge featured a three-way data split: training, leaderboard, and test sets. The training set is publicly available to all participants, the leaderboard data support a leaderboard throughout the competition, and the test set is used to determine a final winner.

Despite significant participation from hundreds of researchers submitting thousands of models over the course of five months, the outcome of the prediction challenge was disappointing. Even the winning model performed hardly better than a simple baseline and predicted little more than the average outcome value.

What caused the poor performance of machine learning on the fragile families data? There are a number of technical possibilities, the sample size, the study design, the missing values. But there is also a more fundamental reason that remains plausible. Perhaps the dynamics of life trajectories are inherently unpredictable over the six year time delay between measurement of the covariates and measurement of the outcome. Machine learning works best in a static and stable world where the past looks like the future. Prediction alone can be a poor choice when we're anticipating dynamic changes, or when we are trying to reason about the effect that hypothetical actions would have in the real world. In subsequent chapters, we will develop powerful conceptual tools to engage more deeply with this observation.

Chapter notes

This chapter overlaps significantly with a chapter on datasets and measurement in the context of fairness and machine learning in the book by Barocas, Hardt, and Narayanan.[19]

The study of adaptivity in data reuse was subject of work by Dwork, Hardt, Pitassi, Reingold, and Roth,[171,172] showing how tools from differential privacy lead to statistical guarantees under adaptivity. Much subsequent work in the area of adaptive data analysis extended these works. A concern closely related to adaptivity goes under the name of *inference after selection* in the statistics community, where it was recognized by Freedman in a 1983 paper.[173]

The notion of leaderboard error and the Ladder algorithm comes from a work by Blum and Hardt.[174] The replication study for CIFAR-10 and ImageNet is due to Recht, Roelofs, Schmidt, and Shankar.[166]

The collection and use of large ad hoc datasets (once referred to as "big data") has been scrutinized in several important works, especially from critical scholars, historians, and social scientists outside the computer science community. See, for example, boyd and Crawford,[175] Tufekci,[176,177] and Onuoha.[178] An excellent survey by Paullada, Raji, Bender, Denton, and Hanna provides a wealth of additional background and references.[179] Olteanu, Castillo, Diaz, and Kiciman discuss biases, methodological pitfalls, and ethical questions in the context of social data analysis.[180] In particular, the article provides comprehensive taxonomies of biases and issues that can arise in the sourcing, collection, processing, and analysis of social data. Recently, Couldry and Mejias use the term *data colonialism* to emphasize the processes by which data are appropriated and marginalized communities are exploited through data collection.[181]

For an introduction to measurement theory, not specific to the social sciences, see the books by Hand.[47,48] The comprehensive textbook by Bandalos[49] focuses on applications to the social sciences, including a chapter on fairness.

Chapter 9

Causality

Our starting point is the difference between an observation and an action. What we see in passive observation is how individuals follow their routine behavior, habits, and natural inclination. Passive observation reflects the state of the world projected to a set of features we choose to highlight. Data that we collect from passive observation show a snapshot of our world as it is.

There are many questions we can answer from passive observation alone. Do 16-year-old drivers have a higher incidence rate of traffic accidents than 18-year-old drivers? Formally, the answer corresponds to a difference of conditional probabilities. We can calculate the conditional probability of a traffic accident given that the driver's age is 16 years and subtract from it the conditional probability of a traffic accident given the age is 18 years. Both conditional probabilities can be estimated from a large enough sample drawn from the distribution, assuming that there are both 16-year-old and 18-year-old drivers. The answer to the question we asked is solidly in the realm of observational statistics.

But important questions often are not observational in nature. Would traffic fatalities decrease if we raised the legal driving age by two years? Although the question seems similar on the surface, we quickly realize that it asks for a fundamentally different insight. Rather than asking for the frequency of an event in our manifested world, this question asks for the effect of a hypothetical action.

As a result, the answer is not so simple. Even if older drivers have a lower incidence rate of traffic accidents, this might simply be a consequence of additional driving experience. There is no obvious reason why an 18-year-old with two months on the road would be any less likely to be involved in an accident than, say, a 16-year-old with the same experience. We can try to address this problem by holding the number of months of driving experience fixed, while comparing individuals of different ages. But we quickly run into subtleties. What if 18-year-olds with two months of driving experience correspond to individuals who are exceptionally cautious and hence—by their natural inclination—not only drive less, but also more cautiously? What if such individuals predominantly live in regions where traffic conditions differ significantly from those in areas where people feel a greater need to drive at a younger age?

We can think of numerous other strategies to answer the original question of whether raising the legal driving age reduces traffic accidents. We could compare countries with different legal driving ages, say, the United States and Germany. But again, these countries differ in many other possibly relevant ways, such as the legal drinking age.

At the outset, causal reasoning is a conceptual and technical framework for addressing questions about the effect of hypothetical actions or *interventions*. Once we understand what the effect of an action is, we can turn the question around and ask what action plausibly *caused* an event. This gives us a formal language to talk about cause and effect.

The limitations of observation

Before we develop any new formalism, it is important to understand why we need it in the first place.

To see why, we turn to the venerable example of graduate admissions at the University of California, Berkeley in 1973.[182] Historical data show that 12,763 applicants were considered for admission to one of 101 departments and interdepartmental majors. Of the 4,321 women who applied, roughly 35 percent were admitted, while 44 percent of the 8,442 men who applied were admitted. Standard statistical significance tests suggest that the observed difference would be highly unlikely to be the outcome of sample fluctuation if there were no difference in underlying acceptance rates.

Table 9.1: UC Berkeley admissions data from 1973

	Men		Women	
Department	Applied	Admitted (%)	Applied	Admitted (%)
A	825	62	108	**82**
B	520	60	25	**68**
C	325	**37**	593	34
D	417	33	375	**35**
E	191	**28**	393	24
F	373	6	341	**7**

A similar pattern exists if we look at the aggregate admission decisions of the six largest departments. The acceptance rate across all six departments for men is about 44%, while it is only roughly 30% for women, again, a significant difference. Recognizing that departments have autonomy over whom to admit, we can look at the gender bias of each department.

What we can see from the table is that four of the six largest departments show a higher acceptance ratio among women, while two show a higher acceptance rate for men. However, these two departments cannot account for the large difference in acceptance rates that we observed in aggregate. So, it appears that the higher acceptance rate for men that we observed in aggregate seems to have reversed at the department level.

Such reversals are sometimes called *Simpson's paradox*, even though mathematically they are no surprise. It's a fact of conditional probability that there can be events Y (here, acceptance), A (here, female gender taken to be a binary variable), and a random variable Z (here, department choice) such that:

1. $\mathbb{P}[Y \mid A] < \mathbb{P}[Y \mid \neg A]$
2. $\mathbb{P}[Y \mid A, Z = z] > \mathbb{P}[Y \mid \neg A, Z = z]$ for all values z that the random variable Z assumes.

Simpson's paradox nonetheless causes discomfort to some, because intuition suggests that a trend that holds for all subpopulations should also hold at the population level.

The reason why Simpson's paradox is relevant to our discussion is that it's a consequence of how we tend to misinterpret what information conditional probabilities encode. Recall that a statement of conditional probability corresponds to passive observation. What we see here is a snapshot of the normal behavior of women and men applying to graduate school at UC Berkeley in 1973.

What is evident from the data is that gender influences department choice. Women and men appear to have different preferences for different fields of study. Moreover, different departments have different admission criteria. Some have lower acceptance rates, some higher. Therefore, one explanation for the data we see is that women *chose* to apply to more competitive departments, hence getting rejected at a higher rate than men.

Indeed, this is the conclusion the original study drew:

> The bias in the aggregated data stems not from any pattern of discrimination on the part of admissions committees, which seems quite fair on the whole, but apparently from prior screening at earlier levels of the educational system. Women are shunted by their socialization and education toward fields of graduate study that are generally more crowded, less productive of completed degrees, and less well funded, and that frequently offer poorer professional employment prospects.[182]

In other words, the article concluded that the source of gender bias in admissions was a *pipeline problem*: Without any wrongdoing by the departments, women were "shunted by their socialization" that happened at an earlier stage in their lives.

It is difficult to debate this conclusion on the basis of the available data alone. The question of discrimination, however, is far from resolved. We can ask why women applied to more competitive departments in the first place. There are several possible reasons. Perhaps less competitive departments, such as engineering schools, were unwelcoming of women at the time. This may have been a general pattern at the time or specific to the university. Perhaps some departments had a track record of poor treatment of women that was known to the applicants. Perhaps the department advertised the program in a manner that discouraged women from applying.

The data we have also shows no measurement of *qualification* of an applicant. It's possible that due to self-selection, women applying to engineering schools in 1973 were overqualified relative to their peers. In this case, an equal acceptance rate between men and women might actually be a sign of discrimination.

There is no way of knowing what was the case from the data we have. We see that at best the original analysis leads to a number of follow-up questions.

At this point, we have two choices. One is to design a new study and collect more data in a manner that might lead to a more conclusive outcome. The other is to argue over which scenario is more likely based on our beliefs and plausible assumptions about the world.

Causal inference is helpful in either case. On the one hand, it can be used as a guide in the design of new studies. It can help us choose which variables to include, which to exclude, and which to hold constant. On the other hand, causal models can serve as a mechanism to incorporate scientific domain knowledge and exchange plausible assumptions for plausible conclusions.

Causal models

We choose *structural causal models* as the basis of our formal discussion as they have the advantage of giving a sound foundation for various causal notions we will encounter. The easiest way to conceptualize a structural causal model is as a program for generating a distribution from independent noise variables through a sequence of formal instructions. Imagine that, instead of samples from a distribution, somebody gave you a step-by-step computer program to generate samples on your own starting from a random seed. The process is not unlike how you would write code. You start from a simple random seed and build up increasingly more complex constructs. That is basically what a structural causal model is, except that each assignment uses the language of mathematics rather than any concrete programming syntax.

A first example

Let's start with a toy example not intended to capture the real world. Imagine a hypothetical population in which an individual exercises regularly with probability 1/2. With probability 1/3, the individual has a latent disposition to become overweight that manifests in the absence of regular exercise. Similarly, in the absence of exercise, heart disease occurs with probability 1/3. Denote by X the indicator variable of regular exercise, by W that of excessive weight, and by H the indicator of heart disease. Below is a structural causal model to generate samples from this hypothetical population. Recall that a Bernoulli random variable $B(p)$ with bias p is a biased coin toss that assumes value 1 with probability p and value 0 with probability $1 - p$.

1. Sample independent Bernoulli random variables $U_1 \sim B(1/2)$, $U_2 \sim B(1/3)$, $U_3 \sim B(1/3)$.
2. $X := U_1$
3. $W :=$ if $X = 1$ then 0 else U_2
4. $H :=$ if $X = 1$ then 0 else U_3

Contrast this generative description of the population with a usual random sample drawn from the population that might look like this:

X	W	H
0	1	1
1	0	0
1	1	1
1	1	0
0	1	0
...	...	...

From the program description, we can immediately see that in our hypothetical population *exercise* averts both *overweight* and *heart disease*, but in the absence of exercise the two are independent. At the outset, our program generates a joint distribution over the random variables (X, W, H). We can calculate probabilities under this distribution. For example, the probability of heart disease under the distribution specified by our model is $1/2 \cdot 1/3 = 1/6$. We can also calculate the conditional probability of heart diseases given overweight. From the event $W = 1$ we can infer that the individual does not exercise, so the probability of heart disease given overweight increases to 1/3 compared with the baseline of 1/6.

Does this mean that overweight causes heart disease in our model? The answer is *no*, as is intuitive given the program to generate the distribution. But let's see how we would go about arguing this point formally. Having a program to generate a distribution is substantially more powerful than just having sampling access. One reason is that we can manipulate the program in whatever way we want, assuming we still end up with a valid program. We could, for example, set $W := 1$, resulting in a new distribution. The resulting program looks like this:

2. $X := U_1$
3. $W := 1$
4. $H :=$ if $X = 1$ then 0 else U_3

This new program specifies a new distribution. We can again calculate the probability of heart disease under this new distribution. We still get $1/6$. This simple calculation reveals a significant insight. The substitution $W := 1$ does not correspond to a conditioning on $W = 1$. One is an action, albeit inconsequential in this case. The other is an observation from which we can draw inferences. If we observe that an individual is overweight, we can infer that they have a higher risk of heart disease (in our toy example). However, this does not mean that lowering body weight would prevent heart disease. It wouldn't in our example. The active substitution $W := 1$ in contrast creates a new hypothetical population in which all individuals are overweight, with all that it entails in our model.

Let us belabor this point a bit more by considering another hypothetical population, specified by the equations:

2. $W := U_2$
3. $X :=$ if $W = 0$ then 0 else U_1
4. $H :=$ if $X = 1$ then 0 else U_3

In this population, exercise habits are driven by body weight. Overweight individuals choose to exercise with some probability, but that's the only reason anyone would exercise. Heart disease develops in the absence of exercise. The substitution $W := 1$ in this model leads to an increased probability of exercise, hence lowering the probability of heart disease. In this case, the conditioning on $W = 1$ has the same affect. Both lead to a probability of $1/6$.

What we see is that fixing a variable by substitution may or may not correspond to a conditional probability. This is a formal rendering of our earlier point that observation isn't action. A substitution corresponds to an action we perform. By substituting a value we break the natural course of action our model captures. This is the reason why the substitution operation is sometimes called the *do-operator*, written as $\mathrm{do}(W := 1)$.

Structural causal models give us a formal calculus to reason about the effect of hypothetical actions. We will see how this creates a formal basis for all the different causal notions that we will encounter in this chapter.

Structural causal models, more formally

Formally, a structural causal model is a sequence of assignments for generating a joint distribution starting from independent noise variables. By executing the sequence of assignments we incrementally build a set of jointly distributed random variables. A structural causal model therefore not only provides a joint distribution, but also a description of how the joint distribution can be generated from elementary noise variables. The formal definition is a bit cumbersome compared with the intuitive notion.

Definition 13. *A* structural causal model M *is given by a set of variables* $X_1, ..., X_d$ *and corresponding assignments of the form*

$$X_i := f_i(P_i, U_i), \qquad i = 1, ..., d\,.$$

Here, $P_i \subseteq \{X_1, ..., X_d\}$ *is a subset of the variables that we call the* parents *of* X_i*. The random variables* $U_1, ..., U_d$ *are called* noise variables, *which we require to be jointly independent.*

The directed graph corresponding to the model has one node for each variable X_i*, which has incoming edges from all the parents* P_i*. We will call such a graph the* causal graph *corresponding to the structural causal model.*

The noise variables that appear in the definition model *exogenous factors* that influence the system. Consider, for example, how the weather influences the delay on a traffic route you choose. Due to the difficulty of modeling the influence of weather more precisely, we could take the weather induced to delay to be an exogenous factor that enters the model as a noise variable. The choice of exogenous variables and their distribution can have important consequences for what conclusions we draw from a model.

The parent nodes P_i of node i in a structural causal model are often called the *direct causes* of X_i. Similarly, we call X_i the *direct effect* of its direct causes P_i. Recall our hypothetical population, in which weight gain was determined by lack of exercise via the assignment $W := \min\{U_1, 1 - X\}$. Here we would say that exercise (or lack thereof) is a direct cause of weight gain.

Structural causal models are a collection of formal *assumptions* about how certain variables interact. Each assignment specifies a *response function*. We can think of nodes as receiving messages from their parents and acting according to these messages as well as the influence of an exogenous noise variable.

To what extent a structural causal model conforms to reality is a separate and difficult question that we will return to in more detail later. For now, think of a structural causal model as formalizing and exposing a set of assumptions about a data generating process. As such, different models can expose different hypothetical scenarios and serve as a basis for discussion. When we make statements about cause and effect in reference to a model, we don't mean to suggest that these relationship necessarily hold in the real world. Whether they

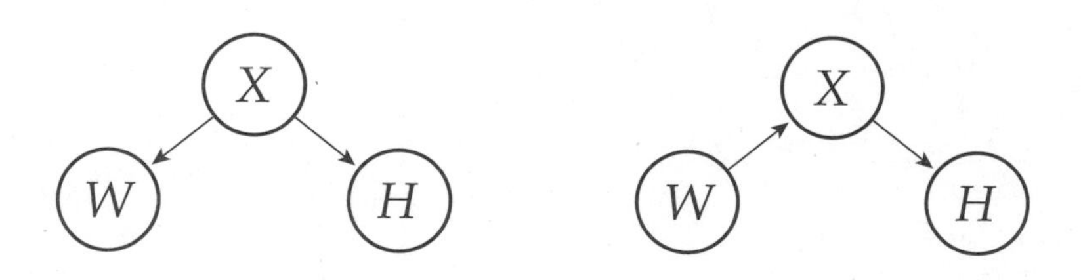

Figure 9.1: Causal diagrams for the heart disease examples

do depends on the scope, purpose, and validity of our model, which may be difficult to substantiate.

It's not hard to show that a structural causal model defines a unique joint distribution over the variables $(X_1, ..., X_d)$ such that $X_i = f_i(P_i, U_i)$. It's convenient to introduce a notion for probabilities under this distribution. When M denotes a structural causal model, we will write the probability of an event E under the entailed joint distribution as $\mathbb{P}_M[E]$. To gain familiarity with the notation, let M denote the structural causal model for the hypothetical population in which both weight gain and heart disease are directly caused by an absence of exercise. We calculated earlier that the probability of heart disease in this model is $\mathbb{P}_M[H] = 1/6$.

In what follows we will derive from this single definition of a structural causal model all the different notions and terminology that we'll need in this chapter.

Throughout, we restrict our attention to acyclic assignments. Many real-world systems are naturally described as stateful dynamical systems with feedback loops. For example, often cycles can be broken up by introducing time-dependent variables, such as investments at time 0 grow the economy at time 1, which in turn grows investments at time 2, continuing so forth until some chosen time horizon t. We will return to a deeper dive into dynamical systems and feedback in later chapters.

Causal graphs

We saw how structural causal models naturally give rise to *causal graphs* that represent the assignment structure of the model graphically. We can go the other way as well by simply looking at directed graphs as placeholders for an unspecified structural causal model which has the assignment structure given by the graph. Causal graphs are often called *causal diagrams*. We'll use these terms interchangeably.

Below we see causal graphs for the two hypothetical populations from our heart disease example.

The scenarios differ in the direction of the link between exercise and weight gain.

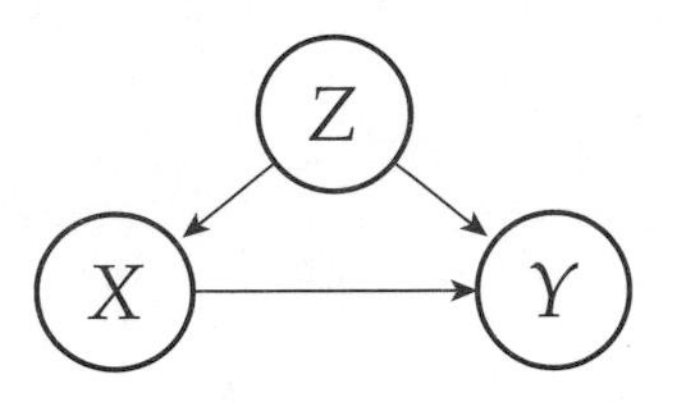

Figure 9.2: Example of a fork

Causal graphs are convenient when the exact assignments in a structural causal model are of secondary importance, but what matters are the paths present and absent in the graph. Graphs also let us import the established language of graph theory to discuss causal notions. We can say, for example, that an *indirect cause* of a node is any ancestor of the node in a given causal graph. In particular, causal graphs allow us to distinguish cause and effect based on whether a node is an ancestor or descendant of another node.

Let's take a first glimpse at a few important graph structures.

Forks

A *fork* is a node Z in a graph that has outgoing edges to two other variables X and Y. Put differently, the node Z is a common cause of X and Y.

We already saw an example of a fork in our weight and exercise example: $W \leftarrow X \rightarrow H$. Here, exercise X influences both weight and heart disease. We also learned from the example that Z has a *confounding* effect: Ignoring exercise X, we saw that W and H appear to be positively correlated. However, the correlation is a mere result of confounding. Once we hold exercise levels constant (via the do-operation), weight has no effect on heart disease in our example.

Confounding leads to a disagreement between the calculus of conditional probabilities (observation) and do-interventions (actions).

Real-world examples of confounding are a common threat to the validity of conclusions drawn from data. For example, in a well-known medical study, a suspected beneficial effect of *hormone replacement therapy* in reducing cardiovascular disease disappeared after identifying *socioeconomic status* as a confounding variable.[183]

Mediators

The case of a fork is quite different from the situation where Z lies on a directed path from X to Y:

In this case, the path $X \rightarrow Z \rightarrow Y$ contributes to the total effect of X on Y. It's a causal path and thus one of the ways in which X causally influences Y. That's why Z is not a confounder. We call Z a *mediator* instead.

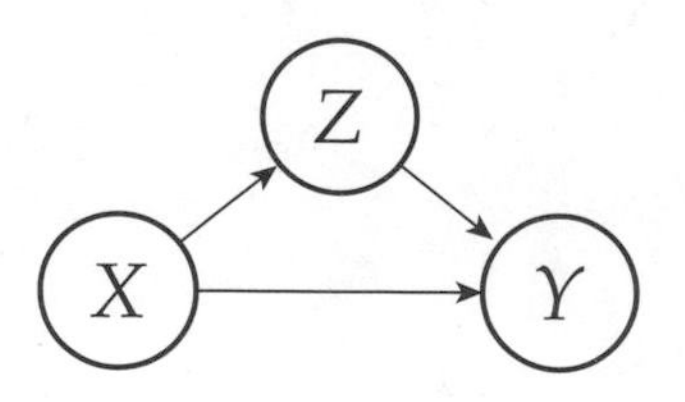

Figure 9.3: Example of a mediator

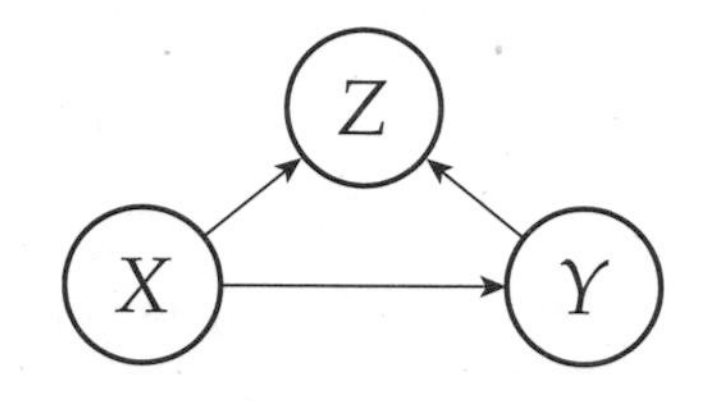

Figure 9.4: Example of a collider

We saw a plausible example of a mediator in our UC Berkeley admissions example. In one plausible causal graph, department choice mediates the influences of gender on the admissions decision.

Colliders

Finally, let's consider another common situation: the case of a *collider*.

Colliders aren't confounders. In fact, in the above graph, X and Y are unconfounded, meaning that we can replace do-statements by conditional probabilities. However, something interesting happens when we condition on a collider. The conditioning step can create correlation between X and Y, a phenomenon called *explaining away*. A good example of the explaining away effect, or *collider bias*, is known as Berkson's paradox.[184] Two independent diseases can become negatively correlated when analyzing hospitalized patients. The reason is that when either disease (X or Y) is sufficient for admission to the hospital (indicated by variable Z), observing that a patient has one disease makes the other statistically less likely. Berkson's paradox is a cautionary tale for statistical analysis when we're studying a cohort that has been subjected to a selection rule.

Interventions and causal effects

Structural causal models give us a way to formalize the effect of hypothetical actions or interventions on the population within the assumptions of our model. As we saw earlier, all we needed was the ability to do substitutions.

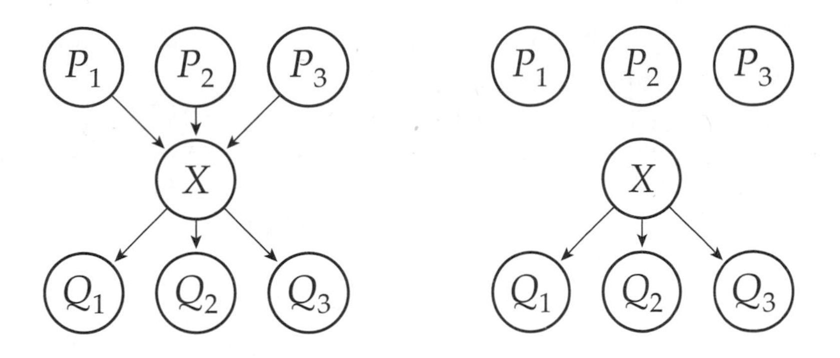

Figure 9.5: Graph before and after substitution

Substitutions and the do-operator

Given a structural causal model M we can take any assignment of the form

$$X := f(P, U)$$

and replace it by another assignment. The most common substitution is to assign X a constant value x:

$$X := x\,.$$

We will denote the resulting model by $M' = M[X := x]$ to indicate the surgery we performed on the original model M. Under this assignment we hold X constant by removing the influence of its parent nodes and thereby any other variables in the model.

Graphically, the operation corresponds to eliminating all incoming edges to the node X. The children of X in the graph now receive a fixed message x from X when they query the node's value.

The assignment operator is also called the *do-operator* to emphasize that it corresponds to performing an action or intervention. We already have notation to compute probabilities after applying the do-operator, namely, $\mathbb{P}_{M[X:=x]}[E]$.

Another notation is popular and common:

$$\mathbb{P}[E \mid \mathrm{do}(X := x)] = \mathop{\mathbb{P}}_{M[X:=x]}[E]\,.$$

This notation analogizes the do-operation with the usual notation for conditional probabilities, and is often convenient when doing calculations involving the do-operator. Keep in mind, however, that the do-operator (action) is fundamentally different from the conditioning operator (observation).

Causal effects

The *causal effect* of an action $X := x$ on a variable Y refers to the distribution of the variable Y in the model $M[X := x]$. When we speak of the causal effect of a variable X on another variable Y we refer to all the ways in which setting X to any possible value x affects the distribution of Y.

Oftentimes X denotes the presence or absence of an intervention or *treatment*. In such cases, X is a binary variable and we are interested in a quantity such as

$$\mathop{\mathbb{E}}_{M[X:=1]}[Y] - \mathop{\mathbb{E}}_{M[X:=0]}[Y] .$$

This quantity is called the *average treatment effect*. It tells us how much treatment (action $X := 1$) increases the expectation of Y relative to no treatment (action $X := 0$).

Causal effects are population quantities. They refer to effects averaged over the whole population. Often the effect of treatment varies greatly from one individual or group of individuals to another. Such treatment effects are called *heterogeneous*.

Confounding

Important questions in causality relate to when we can rewrite a do-operation in terms of conditional probabilities. When this is possible, we can estimate the effect of the do-operation from conventional conditional probabilities that we can estimate from data.

The simplest question of this kind asks when a causal effect $\mathbb{P}[Y = y \mid \mathrm{do}(X := x)]$ coincides with the condition probability $\mathbb{P}[Y = y \mid X = x]$. In general, this is not true. After all, the difference between observation (conditional probability) and action (interventional calculus) is what motivated the development of causality.

The disagreement between interventional statements and conditional statements is so important that it has a well-known name: *confounding*. We say that X and Y are confounded when the causal effect of action $X := x$ on Y does not coincide with the corresponding conditional probability.

When X and Y are confounded, we can ask if there is some combination of conditional probability statements that give us the desired effect of a do-intervention. This is generally possible given a causal graph by conditioning on the parent nodes PA of the node X:

$$\mathbb{P}[Y = y \mid \mathrm{do}(X := x)] = \sum_z \mathbb{P}[Y = y \mid X = x, PA = z] \, \mathbb{P}[PA = z] .$$

This formula is called the *adjustment formula*. It gives us one way of estimating the effect of a do-intervention in terms of conditional probabilities.

The adjustment formula is one example of what is often called *controlling for* a set of variables. We estimate the effect of X on Y separately in every slice of the population defined by a condition $Z = z$ for every possible value of z. We then average these estimated sub-population effects weighted by the probability of $Z = z$ in the population. To give an example, when we control for age, we mean that we estimate an effect separately in each possible age group and then average out the results so that each age group is weighted by the fraction of the population that falls into the age group.

Controlling for more variables in a study isn't always the right choice. It depends on the graph structure. Let's consider what happens when we control for the variable Z in the three causal graphs we discussed above.

- Controlling for a confounding variable Z in a fork $X \leftarrow Z \rightarrow Y$ will deconfound the effect of X on Y.
- Controlling for a mediator Z will eliminate some of the causal influence of X on Y.
- Controlling for a collider will create correlation between X and Y. That is the opposite of what controlling for Z accomplishes in the case of a fork. The same is true if we control for a descendant of a collider.

The backdoor criterion

At this point, we might worry that things will get increasingly complicated. As we introduce more nodes in our graph, we might fear a combinatorial explosion of possible scenarios to discuss. Fortunately, there are simple sufficient criteria for choosing a set of deconfounding variables that is safe to control for.

A well-known graph-theoretic notion is the *backdoor* criterion.[185] Two variables are confounded if there is a so-called *backdoor* path between them. A *backdoor path* from X to Y is any path starting at X with a backward edge "$\leftarrow$" into X, such as:

$$X \leftarrow A \rightarrow B \leftarrow C \rightarrow Y$$

Intuitively, backdoor paths allow information flow from X to Y in a way that is not causal. To deconfound a pair of variables we need to select a *backdoor set* of variables that "blocks" all backdoor paths between the two nodes. A backdoor path involving a chain $A \rightarrow B \rightarrow C$ can be blocked by controlling for B. Information by default cannot flow through a collider $A \rightarrow B \leftarrow C$. So we only have to be careful not to open information flow through a collider by conditioning on the collider, or a descendant of a collider.

Unobserved confounding

The adjustment formula might suggest that we can always eliminate confounding bias by conditioning on the parent nodes. However, this is only true in the absence of *unobserved confounding*. In practice, often there are variables that are hard to measure, or were simply left unrecorded. We can still include such

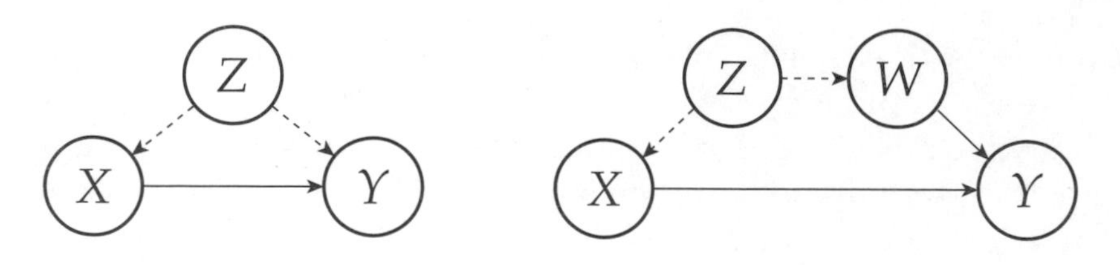

Figure 9.6: Two cases of unobserved confounding

unobserved nodes in a graph, typically denoting their influence with dashed lines, instead of solid lines.

The above figure shows two cases of unobserved confounding. In the first example, the causal effect of X on Y is unidentifiable. In the second case, we can block the confounding backdoor path $X \leftarrow Z \rightarrow W \rightarrow Y$ by controlling for W even though Z is not observed. The backdoor criterion lets us work around unobserved confounders in some cases where the adjustment formula alone wouldn't suffice.

Unobserved confounding nonetheless remains a major obstacle in practice. The issue is not just lack of measurement, but often lack of anticipation or awareness of a confounding variable. We can try to combat unobserved confounding by increasing the number of variables under consideration. But as we introduce more variables into our study, we also increase the burden of coming up with a valid causal model for all variables under consideration. In practice, it is not uncommon to control for as many variables as possible with the hope of disabling confounding bias. However, as we saw, controlling for mediators or colliders can be harmful.

Randomization and the backdoor criterion

The backdoor criterion gives a non-experimental way of eliminating confounding bias given a causal model and a sufficient amount of observational data from the joint distribution of the variables. An alternative experimental method of eliminating confounding bias is randomization.

The idea is simple. If a treatment variable T is an unbiased coin toss, nothing but mere chance influenced its assignment. In particular, there cannot be a confounding variable exercising influence on both the treatment variable and a desired outcome variable.

A different way to think about this is that randomization breaks natural inclination. Rather than letting treatment take on its natural value, we assign it randomly. Thinking in terms of causal models, what this means is that we eliminate all incoming edges into the treatment variable. In particular, this closes all backdoor paths and hence avoids confounding bias. Because randomization is such an important part of causal inference, we now turn to it in greater detail.

Experimentation, randomization, potential outcomes

Let's think about experimentation from first principles. Suppose we have a population of individuals and we have devised some treatment that can be applied to each individual. We would like to know the effect of this treatment on some measurable quantity.

As a simple example, and one which has had great utility, consider the development of a vaccine for a disease. How can we tell if a vaccine prevents disease? If we give everyone the vaccine, we'd never be able to disentangle whether the treatment caused the associated change in disease we observe or not. The common and widely accepted solution in medicine is to restrict our attention to a subset of the population, and leverage randomized assignment to isolate the effect of the treatment.

The simplest mathematical formulation underlying randomized experiment design is as follows. We assume a group of n individuals $i = 1, \ldots, n$. Suppose for an individual that if we apply a treatment, the quantity of interest is equal to a value $Y_1(i)$. If we don't apply the treatment, the quantity of interest is equal to $Y_0(i)$. We define an outcome $Y(i)$ that is equal to $Y_1(i)$ if the treatment is applied and equal to $Y_0(i)$ if the treatment is not applied. In our vaccine example, $Y_1(i)$ indicates whether a person contracts the disease in a specified time period following a vaccination and $Y_0(i)$ indicates whether a person contracts the disease in the same time period absent vaccination. Now, obviously, one person can only take one of these paths. Nonetheless, we can imagine two *potential outcomes*: one potential outcome $Y_1(i)$ if the treatment is applied and another potential outcome $Y_0(i)$ if the treatment is not applied. Throughout this section, we assume that the potential outcomes are fixed deterministic values.

We can write the relationship between observed outcome and potential outcomes as a mathematical equation by introducing the Boolean treatment indicator $T(i)$, which is equal to 1 if subject i receives the treatment and 0 otherwise. In this case, the outcome for individual i equals

$$Y(i) = T(i)Y_1(i) + (1 - T(i))Y_0(i) .$$

That is, if the treatment is applied, we observe $Y_1(i)$, and if the treatment is not applied, we observe $Y_0(i)$. While this potential outcome formulation is tautological, it lets us apply the same techniques we use for estimating the mean to the problem of estimating treatment effects.

The individual treatment effect is a relation between the quantities $Y_1(i)$ and $Y_0(i)$, commonly just the difference $Y_1(i) - Y_0(i)$. If the difference is positive, we see that applying the treatment increases the outcome variable for this individual. But, as we've discussed, our main issue is that we can never simultaneously observe $Y_1(i)$ and $Y_0(i)$. Once we choose whether to apply the treatment or not, we can only measure the corresponding treated or untreated condition.

While it may be daunting to predict the treatment effect at the level of each individual, statistical algorithms can be applied to estimate average treatment effects across the population. There are many ways to define a measure of the effect of a treatment on a population. For example, we earlier defined the notion of an average treatment effect. Let $\bar{Y}_1$ and $\bar{Y}_0$ denote the means of $Y_1(i)$ and $Y_0(i)$, respectively, averaged over $i = 1, \ldots, n$. We can write

$$\text{Average Treatment Effect} = \bar{Y}_1 - \bar{Y}_0 .$$

In the vaccine example, this would be the difference in the probability of contracting the illness if one were vaccinated versus if one were not vaccinated.

The *odds* that an individual catches the disease is the number of people who catch the disease divided by the number who do not. The *odds ratio* for a treatment is the odds when every person receives the vaccine divided by the odds when no one receives the vaccine. We can write this out as a formula in terms of our quantities $\bar{Y}_1$ and $\bar{Y}_0$. When the potential outcomes take on values 0 or 1, the average $\bar{Y}_1$ is the number of individuals for which $Y_1(i) = 1$ divided by the total number of individuals. Hence, we can write the odds ratio as

$$\text{Odds Ratio} = \frac{\bar{Y}_1}{1 - \bar{Y}_1} \cdot \frac{1 - \bar{Y}_0}{\bar{Y}_0} .$$

This measures the decrease (or increase!) of the odds of a bad event happening when the treatment is applied. When the odds ratio is less than 1, the odds of a bad event are lower if the treatment is applied. When the odds ratio is greater than 1, the odds of a bad event are higher if the treatment is applied.

Similarly, the *risk* that an individual catches the disease is the ratio of the number of people who catch the disease to the total population size. Risk and odds are similar quantities, but some disciplines prefer one to the other by convention. The *risk ratio* is the fraction of bad events when a treatment is applied divided by the fraction of bad events when it is not applied:

$$\text{Risk Ratio} = \frac{\bar{Y}_1}{\bar{Y}_0} .$$

The risk ratio measures the increase or decrease of relative risk of a bad event when the treatment is applied. In the context of vaccines, this ratio is popularly reported differently. The *effectiveness* of a treatment is 1 minus the risk ratio. This is precisely the number used when people say a vaccine is 95% effective. It is equivalent to saying that the proportion of those treated who fell ill was 20 times less than the proportion of those not treated who fell ill. Importantly, it does not mean that one has a 5% chance of contracting the disease.

Estimating treatment effects using randomization

Let's now analyze how to estimate these effects using a randomized procedure. In a randomized controlled trial a group of n subjects is randomly partitioned into a *control group* and a *treatment group*. We assume participants do not know which group they were assigned to and neither does the staff administering the trial. The treatment group receives an actual treatment, such as a drug that is being tested for efficacy, while the control group receives a placebo identical in appearance. An outcome variable is measured for all subjects.

Formally, this means each $T(i)$ is an unbiased coin toss. Because we randomly assign treatments, we have

$$\mathbb{E}[Y(i) \mid T(i) = 1] = Y_1(i) \qquad \text{and} \qquad \mathbb{E}[Y(i) \mid T(i) = 0] = Y_0(i)\,.$$

Therefore, for treatment value $t \in \{0, 1\}$,

$$\mathbb{E}\left[\frac{1}{n}\sum_{i=1}^{n} Y(i) \mid T(i) = t\right] = \bar{Y}_t\,.$$

In other words, to get an unbiased estimate of $\bar{Y}_t$ we just have to average out all outcomes for subjects with treatment assignment t. This in turn gives us the various causal effects we discussed previously. We can also apply more statistics to this estimate to get confidence bounds and large deviation bounds. Various things we know for estimating the mean of a population carry over. For example, we need the outcome variables to have bounded range in order for our estimates to have low variance. Similarly, if we are trying to detect a tiny causal effect, we must choose n sufficiently large.

Typically in a randomized control trial, the n subjects are supposed to a uniformly random sample from a larger target population of N individuals. The group average $\bar{Y}_t$ is therefore itself only an estimate of the population mean. Here, too, conventional statistics applies in reasoning about how close $\bar{Y}_t$ is to the population average.

Uniform sampling from a population is an idealization that is hard to achieve in experimental practice. It is not only hard to independently sample individuals in a large population, but we also need to be able to set up identical scenarios to test interventions. For medical treatments, what if there is variance between the treatment effect at 9 AM in the Mayo Clinic on a Tuesday and at 11 PM in the Alta Bates Medical Center on a Saturday? If there are temporal or spatial or other variabilities, the effective size of the population and the corresponding variance grow. Accounting for such variability is a daunting challenge of modern medical and social research that can be at the root of many failures of replication.

The formulation here also assumes that the potential outcomes $Y_t(i)$ do not vary over time. The framework could be generalized to account for temporal variation, but such a generalization will not illuminate the basic issues of statistical methods and modeling. We return to the practice of causal inference and its challenges in the next chapter. But before we do so, we will relate what we just learned to the structural causal models that we saw earlier.

Counterfactuals

Fully specified structural causal models allow us to ask causal questions that are more delicate than the mere effect of an action. Specifically, we can ask *counterfactual* questions such as: Would I have avoided the traffic jam had I taken a different route this morning?

Formally, counterfactuals are random variables that generalize the potential outcome variables we saw previously. Counterfactuals derive from a structural causal model, which gives as another useful way to think about potential outcomes. The procedure for extracting counterfactuals from a structural causal model is algorithmic, but it can look a bit subtle at first. It helps to start with a simple example.

A simple counterfactual

Assume every morning we need to decide between two routes $T = 0$ and $T = 1$. On bad traffic days, indicated by $U = 1$, both routes are bad. On good days, indicated by $U = 0$, the traffic on either route is good unless there was an accident on the route.

Let's say that $U \sim B(1/2)$ follows the distribution of an unbiased coin toss. Accidents occur independently on either route with probability 1/2. So, choose two Bernoulli random variables $U_0, U_1 \sim B(1/2)$ that tell us if there is an accident on route 0 and route 1, respectively. We reject all external route guidance and instead decide on which route to take uniformly at random. That is, $T := U_T \sim B(1/2)$ is also an unbiased coin toss.

Introduce a variable $Y \in \{0,1\}$ that tells us whether the traffic on the chosen route is good ($Y = 0$) or bad ($Y = 1$). Reflecting our discussion above, we can express Y as

$$Y := T \cdot \max\{U, U_1\} + (1 - T) \max\{U, U_0\} .$$

In words, when $T = 0$ the first term disappears and so traffic is determined by the larger of the two values U and U_0. Similarly, when $T = 1$ traffic is determined by the larger of U and U_1.

Now, suppose one morning we have $T = 1$ and we observe bad traffic $Y = 1$. Would we have been better off taking the alternative route this morning?

A natural attempt to answer this question is to compute the likelihood of $Y = 0$ after the do-operation $T := 0$, that is, $\mathbb{P}_{M[T:=0]}[Y = 0]$. A quick calculation reveals that this probability is $\frac{1}{2} \cdot \frac{1}{2} = 1/4$. Indeed, given the substitu-

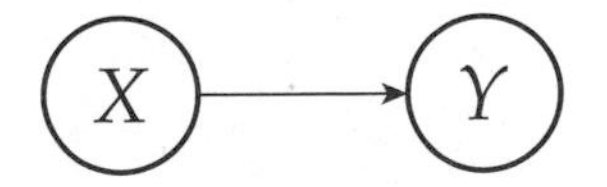

Figure 9.7: Causal diagram for our traffic scenario

tion $T := 0$ in our model, for the traffic to be good we need that $\max\{U, U_0\} = 0$. This can only happen when both $U = 0$ (probability $1/2$) and $U_0 = 0$ (probability $1/2$).

But this isn't the correct answer to our question. The reason is that we took route $T = 1$ and observed that $Y = 1$. From this observation, we can deduce that certain background conditions did not manifest for they are inconsistent with the observed outcome. Formally, this means that certain settings of the noise variables (U, U_0, U_1) are no longer feasible given the observed event $\{Y = 1, T = 1\}$. Specifically, if U and U_1 had both been zero, we would have seen no bad traffic on route $T = 1$, but this is contrary to our observation. In fact, the evidence $\{Y = 1, T = 1\}$ leaves only the following settings for U and U_1:

Table 9.3: Possible noise settings after observing evidence

U	U_1
0	1
1	1
1	0

We leave out U_0 from the table, since its distribution is unaffected by our observation. Each of these three cases is equally likely, which in particular means that the event $U = 1$ now has probability $2/3$. In the absence of any additional evidence, recall $U = 1$ had probability $1/2$. What this means is that the observed evidence $\{Y = 1, T = 1\}$ has biased the distribution of the noise variable U toward 1. Let's use the letter U' to refer to this biased version of U. Formally, U' is distributed according to the distribution of U conditional on the event $\{Y = 1, T = 1\}$.

Working with this biased noise variable, we can again entertain the effect of the action $T := 0$ on the outcome Y. For $Y = 0$ we need that $\max\{U', U_0\} = 0$. This means that $U' = 0$, an event that now has probability $1/3$, and $U_0 = 0$ (probability $1/2$ as before). Hence, we get the probability $1/6 = 1/2 \cdot 1/3$ for the event that $Y = 0$ under our do-operation $T := 0$, and after updating the noise variables to account for the observation $\{Y = 1, T = 1\}$.

To summarize, incorporating available evidence into our calculation decreased the probability of no traffic ($Y = 0$) when choosing route 0 from $1/4$ to $1/6$. The intuitive reason is that the evidence made it more likely that it was generally a bad traffic day, and even the alternative route would've been clogged. The event we observed biases the distribution of exogenous variables.

We think of the result we just calculated as the *counterfactual* of choosing the alternative route given the route we chose had bad traffic.

The general recipe

We can generalize our discussion of computing counterfactuals from the previous example to a general procedure. There were three essential steps. First, we incorporated available observational evidence by biasing the exogenous noise variables through a conditioning operation. Second, we performed a do-operation in the structural causal model after we substituted the biased noise variables. Third, we computed the distribution of a target variable.

These three steps are typically called *abduction*, *action*, and *prediction*, as can be described as follows.

Definition 14. *Given a structural causal model M, an observed event E, an action $T := t$, and a target variable Y, we define the* counterfactual $Y_{T:=t}(E)$ *by the following three-step procedure:*

1. ***Abduction:*** *Adjust noise variables to be consistent with the observed event. Formally, condition the joint distribution of $U = (U_1, ..., U_d)$ on the event E. This results in a biased distribution U'.*
2. ***Action:*** *Perform do-intervention $T := t$ in the structural causal model M, resulting in the model $M' = M[T := t]$.*
3. ***Prediction:*** *Compute target counterfactual $Y_{T:=t}(E)$ by using U' as the random seed in M'.*

It's important to realize that this procedure *defines* what a counterfactual is in a structural causal model. The notation $Y_{T:=t}(E)$ denotes the outcome of the procedure and is part of the definition. We haven't encountered this notation before.

Put in words, we interpret the formal counterfactual $Y_{T:=t}(E)$ as the value Y would've taken had the variable T been set to value t in the circumstances described by the event E.

In general, the counterfactual $Y_{T:=t}(E)$ is a random variable that varies with U'. But counterfactuals can also be deterministic. When the event E narrows down the distribution of U to a single point mass, called *unit*, the variable U' is constant and hence the counterfactual $Y_{T:=t}(E)$ reduces to a single number. In this case, it's common to use the shorthand notation $Y_t(u) = Y_{T:=t}(\{U = u\})$, where we make the variable t implicit, and let u refer to a single unit. The counterfactual random variable Y_t refers to $Y_t(u)$ for a random draw of the noise variables u.

The motivation for the name *unit* derives from the common situation where the structural causal model describes a population of entities that form the atomic units of our study. It's common for a unit to be an individual (or the description of a single individual). However, depending on application, the choice of units can vary. In our traffic example, the noise variables dictate which route we take and what the road conditions are.

Answers to counterfactual questions strongly depend on the specifics of the structural causal model, including the precise model of how the exogenous noise variables come into play. It's possible to construct two models that have identical graph structures, and behave identically under interventions, yet give different answers to counterfactual queries.[186]

Potential outcomes

Let's return to the *potential outcomes* framework that we introduced when discussing randomized experiments. Rather than deriving potential outcomes from a structural causal model, we assume their existence as ordinary random variables, albeit some unobserved. Specifically, we assume that for every unit u there exist random variables $Y_t(u)$ for every possible value of the assignment t. This potential outcome turns out to equal the corresponding counterfactual derived from the structural equation model:

$$\text{potential outcome } Y_t(u) = Y_{T:=t}(\{u\}) \text{ structural counterfactual}\,.$$

In particular, there is no harm in using our potential outcome notation $Y_t(u)$ as a shorthand for the corresponding counterfactual notation.

In the potential outcomes model, it's customary to think of a binary *treatment variable* T that assumes only two values, 0 for *untreated*, and 1 for *treated*. This gives us two potential outcome variables $Y_0(u)$ and $Y_1(u)$ for each unit u. There is some potential for notational confusion here. Readers already familiar with the potential outcomes model may be used to the notation "$Y_i(0), Y_i(1)$" for the two potential outcomes corresponding to unit i. In our notation the unit (or, more generally, set of units) appears in the parentheses and the subscript denotes the substituted value for the variable we intervene on.

The key point about the potential outcomes model is that we only observe the potential outcome $Y_1(u)$ for units that were treated. For untreated units we observe $Y_0(u)$. In other words, we can never simultaneously observe both, although they're both assumed to exist in a formal sense. Formally, the outcome $Y(u)$ for unit u that we observe depends on the binary treatment $T(u)$ and is given by the expression:

$$Y(u) = Y_0(u) \cdot (1 - T(u)) + Y_1(u) \cdot T(u)\,.$$

We can revisit our traffic example in this framework. The next table summarizes what information is observable in the potential outcomes model. We think of the route we choose as the treatment variable, and the observed traffic as reflecting one of the two potential outcomes.

Table 9.4: Traffic example in the potential outcomes model

Route T	Outcome Y_0	Outcome Y_1	Probability
0	0	?	1/8
0	1	?	3/8
1	?	0	1/8
1	?	1	3/8

Often this information comes in the form of samples. For example, we might observe the traffic on different days. With sufficiently many samples, we can estimate the above frequencies with arbitrary accuracy.

Table 9.5: Traffic data in the potential outcomes model

Day	Route T	Outcome Y_0	Outcome Y_1
1	0	1	?
2	0	0	?
3	1	?	1
4	0	1	?
5	1	?	0
...	...	...	...

In our original traffic example, there were 16 units corresponding to the background conditions given by the four binary variables U, U_0, U_1, U_T. When the units in the potential outcome model agree with those of a structural causal model, then causal effects computed in the potential outcomes model agree with those computed in the structural equation model. The two formal frameworks are perfectly consistent with each other.

As is intuitive from the table above, causal inference in the potential outcomes framework can be thought of as filling in the missing entries ("?") in the table above. This is sometimes called *missing data imputation* and there are numerous statistical methods for this task. If we could *reveal* what's behind the question marks, many quantities would be readily computable. For instance, estimating the average treatment effect would be as easy as counting rows.

When we were able to directly randomize the treatment variable, we showed that treatment effects could be imputed from samples. When we are working with observational data, there is a set of established conditions under which causal inference becomes possible:

1. **Stable Unit Treatment Value Assumption** (SUTVA): The treatment that one unit receives does not change the effect of treatment for any other unit.

2. **Consistency**: Formally, $Y(u) = Y_0(u)(1 - T(u)) + Y_1(u)T(u)$. That is, $Y(u) = Y_0(u)$ if $T(u) = 0$ and $Y(u) = Y_1(u)$ if $T(u) = 1$. In words, the outcome $Y(u)$ agrees with the potential outcome corresponding to the treatment indicator.
3. **Ignorability**: The potential outcomes are independent of treatment given some deconfounding variables Z, i.e., $T \perp (Y_0, Y_1) \mid Z$. In words, the potential outcomes are conditionally independent of treatment given some set of deconfounding variables.

The first two assumptions automatically hold for counterfactual variables derived from structural causal models according to the procedure described above. This assumes that the units in the potential outcomes framework correspond to the atomic values of the background variables in the structural causal model.

The third assumption is a major one. The assumption on its own cannot be verified or falsified, since we never have access to samples with both potential outcomes manifested. However, we can verify if the assumption is consistent with a given structural causal model, for example, by checking if the set Z blocks all backdoor paths from treatment T to outcome Y.

There's no tension between structural causal models and potential outcomes and there's no harm in having familiarity with both. It nonetheless makes sense to say a few words about the differences of the two approaches.

We can derive potential outcomes from a structural causal model as we did above, but we cannot derive a structural causal model from potential outcomes alone. A structural causal model in general encodes more assumptions about the relationships of the variables. This has several consequences. On the one hand, a structural causal model gives us a broader set of formal concepts (causal graphs, mediating paths, counterfactuals for every variable, and so on). On the other hand, coming up with a plausibly valid structural causal model is often a daunting task that might require knowledge that is simply not available. Difficulty to come up with a plausible causal model often exposes unsettled substantive questions that require resolution first.

The potential outcomes model, in contrast, is generally easier to apply. There's a broad set of statistical estimators of causal effects that can be readily applied to observational data. But the ease of application can also lead to abuse. The assumptions underpinning the validity of such estimators are experimentally unverifiable. Our next chapter dives deeper into the practice of causal inference and some of its limitations.

Chapter notes

This chapter was developed and first published by Barocas, Hardt, and Narayanan in the textbook *Fairness and Machine Learning: Limitations and Opportunities.*[19]

With permission from the authors, we include a large part of the original text here with only slight modifications. We removed a significant amount of material on discrimination and fairness and added an extended discussion on randomized experiments.

There are several excellent introductory textbooks on the topic of causality. For an introduction to causality with an emphasis on causal graphs and structural equation models, turn to Pearl's primer,[187] or the more comprehensive textbook.[185] Our exposition of Simpson's paradox and the UC Berkeley data was influenced by Pearl's discussion, updated for a new popular audience book.[188] The example has been heavily discussed in various other writings, such as Pearl's recent discussion.[188] We retrieved the Berkeley data from `http://www.randomservices.org/random/data/Berkeley.html`. There is some discrepancy with the data available on the Wikipedia page for Simpson's paradox that we retrieved on December 27, 2018.

For further discussion regarding the popular interpretation of Simpson's original article,[189] see the article by Hernán, Clayton, and Keiding,[190] as well as Pearl's text.[185]

The technically minded reader will enjoy complementing Pearl's book with the recent open access text by Peters, Janzing, and Schölkopf[186] that is available online. The text emphasizes two variable causal models and applications to machine learning. See Spirtes, Glymour, and Scheines[191] for a general introduction based on causal graphs with an emphasis on *graph discovery*, i.e., inferring causal graphs from observational data. An article by Schölkopf provides additional context about the development of causality in machine learning.[192]

The classic formulation of randomized experiment design due to Jerzy Neyman is now subsumed by and commonly referred to as the framework of potential outcomes.[193,194] Imbens and Rubin[195] give a comprehensive introduction to the technical repertoire of causal inference in the potential outcomes model. Angrist and Pischke[196] focus on causal inference and potential outcomes in econometrics. Hernán and Robins[197] give another detailed introduction to causal inference that draws on the authors' experience in epidemiology. Morgan and Winship[198] focus on applications in the social sciences.

Chapter 10

Causal Inference in Practice

The previous chapter introduced the conceptual foundations of causality, but there's a lot more to learn about how these concepts play out in practice. In fact, there's a flourishing practice of causal inference in numerous scientific disciplines. Increasingly, ideas from machine learning show up in the design of causal estimators. Conversely, ideas from causal inference can help machine learning practitioners run better experiments.

In this chapter we focus on estimating the average treatment effect, often abbreviated as ATE, of a binary treatment T on an outcome variable Y:

$$\mathbb{E}[Y \mid \text{do}(T := 1)] - \mathbb{E}[Y \mid \text{do}(T := 0)] .$$

Causal effects are population quantities that involve two hypothetical actions, one holding the treatment variable constant at the treatment value 1, the other holding the treatment constant at its baseline value 0.

The central question in causal inference is how we can estimate causal quantities, such as the average treatment effect, from data.

Confounding between the outcome and treatment variable is the main impediment to causal inference from observational data. Recall that random variables Y and T are confounded if the conditional probability distribution of Y given T does not equal its interventional counterpart:

$$\mathbb{P}[Y = y \mid \text{do}(T := t)] \neq \mathbb{P}[Y = y \mid T = t] .$$

If these expressions were equal, we could estimate the average treatment effect in a direct way by estimating the difference $\mathbb{E}[Y \mid T = 1] - \mathbb{E}[Y \mid T = 0]$ from samples. Confounding makes the estimation of treatment effects more challenging, and sometimes impossible. Note that the main challenge here is to arrive at an expression for the desired causal effect that is free of any causal constructs, such as the do-operator. Once we have a plain probability expression at hand, tools from statistics allow us to relate the population quantity with a finite sample estimate.

Design and inference

There are two important components to causal inference, one is *design*, the other is *inference*.

In short, design is about sorting out various substantive questions about the data generating process. Inference is about the statistical apparatus that we unleash on the data in order to estimate a desired causal effect.

Design requires us to decide on a population, a set of variables to include, and a precise question to ask. In this process we need to engage substantively with relevant scientific domain knowledge in order to understand what assumptions we can make about the data.

Design can only be successful if the assumptions we are able to make permit the estimation of the causal effect we're interested in. In particular, this is where we need to think carefully about potential sources of confounding and how to cope with them.

There is no way statistical estimators can recover from poor design. If the design does not permit causal inference, there is simply no way that a clever statistical trick could remedy the shortcoming. It's therefore apt to think of causal insights as consequences of the substantive assumptions that we can make about the data, rather than as products of sophisticated statistical ideas.

Hence, we emphasize design issues throughout this chapter and intentionally do not dwell on technical statements about rates of estimation. Such mathematical statements can be valuable, but design must take precedence.

Experimental and observational designs

Causal inference distinguishes between *experimental* and *observational* designs. Experimental designs generally are active in the sense of administering some treatment to some set of experimental units. Observational designs do not actively assign treatment, but rather aim to make it possible to identify causal effects from collected data without implementing any interventions.

The most common and well-established experimental design is a randomized controlled trial (RCT). The main idea is to assign treatment randomly. A randomly assigned treatment, by definition, is not influenced by any other variable. Hence, randomization eliminates any confounding bias between treatment and outcome.

In a typical implementation of a randomized controlled trial, subjects are randomly partitioned into a *treatment group* and a *control group*. The treatment group receives the treatment, the control group receives no treatment. It is important that subjects do not know which group they were assigned to. Otherwise knowledge of their assignment may influence the outcome. To ensure this, subjects in the control group receive what is called a *placebo*, a device or procedure that looks indistinguishable from treatment to the study subject, but lacks the treatment ingredient whose causal powers are in question. Adequate placebos may not exist depending on what the treatment is, for example, in the case of a surgery.

Randomized controlled trials have a long history with many success stories. They've become an important source of scientific knowledge.

Sometimes randomized controlled trials are difficult, expensive, or impossible to administer. Treatment might be physically or legally impossible, too costly, or too dangerous. Nor are they free of issues and pitfalls.[199] In this chapter, we will see observational alternatives to randomized controlled trials. However, these are certainly not without their own set of difficulties and shortcomings.

The machine learning practitioner is likely to encounter randomization in the form of so-called *A/B tests*. In an A/B test we randomly assign one of two treatments to a set of individuals. Such experiments are common in the tech industry to find out which of two changes to a product leads to a better outcome.

The observational basics: Adjustment and controls

For the remainder of the chapter we focus on observational causal inference methods. In the previous chapter we saw that there are multiple ways to cope with confounding between treatment and outcome. One of them is to adjust (or control) for the parents (i.e., direct causes) of T via the adjustment formula.

The extra variables that we adjust for are also called *controls*, and we take the phrase *controlling for* to mean the same thing as *adjusting for*.

We then saw that we could use any set of random variables satisfying the graphical backdoor criterion. This is helpful in cases where some direct causes are unobserved so that we cannot use them in the adjustment formula.

Let's generalize this idea even further and call a set of variables *admissible* if it satisfies the adjustment formula.

Definition 15. *We say that a discrete random variable X is* admissible *if it satisfies the adjustment formula:*

$$\mathbb{P}[Y = y \mid \mathrm{do}(T := t)] = \sum_x \mathbb{P}[Y = y \mid T = t, X = x] \, \mathbb{P}[X = x] \, .$$

Here we sum over all values x in the support of X.

The definition directly suggests a basic estimator for the do-intervention.

Basic adjustment estimator.

1. Collect samples n samples $(t_i, y_i, x_i)_{i=1}^n$.
2. Estimate each of the conditional probabilities $\mathbb{P}[Y = y \mid T = t, X = x]$ from the collected samples.
3. Compute the sum $\sum_x \mathbb{P}[Y = y \mid T = t, X = x] \, \mathbb{P}[X = x]$.

This estimator can only work if all slices $\{T = t, X = x\}$ have nonzero probability, an assumption often called *overlap* or *positivity* in causal inference.

But the basic estimator also fails when the adjustment variable X can take on too many possible values. In general, the variable X could correspond to a tuple of features, such as, age, height, weight, etc. The support of X grows exponentially with the number of features. This poses an obvious computational problem, but more importantly a statistical problem as well. By a counting argument some of the events $\{T = t, X = x\}$ must have probability as small as the inverse of size of the support X. To estimate a probability $p > 0$ from samples to within small relative error, we need about $O(1/p^2)$ samples.

Much work in causal inference deals with overcoming the statistical inefficiency of the basic estimator. Conceptually, however, most sophisticated estimators work from the same principle. We need to assume that we have an admissible variable X and that positivity holds. Different estimators then use this assumption in different ways.

Potential outcomes and ignorability

The average treatment effect often appears in the causal inference literature equivalently in its potential outcome notation $\mathbb{E}[Y_1 - Y_0]$. This way of going about it is mathematically equivalent and either way works for us.

When talking about potential outcomes, it's customary to replace the assumption that X is admissible with another essentially equivalent assumption called *ignorability* or *unconfoundedness*. To recall from the previous chapter, this assumption requires that the potential outcomes variables are conditionally independent of treatment given X. Formally, $T \perp (Y_0, Y_1) \mid X$. It's not hard to show that ignorability implies that X is admissible.

Reductions to model fitting

Adjustment gives a simple and general way to estimate causal effects given an admissible set of variables. The primary shortcoming that we discussed is the sample inefficiency of the formula in high-dimensional settings.

There's a vast literature of causal estimators that aim to address this central shortcoming in a range of different settings. While the landscape of causal estimators might seem daunting to newcomers, almost all causal inference methods share a fundamental idea. This idea is reduce causal inference to standard supervised machine learning tasks.

Let's see how this central idea plays out in a few important cases.

Propensity scores

Propensity scores are one popular way to cope with adjustment variables that have large support. Let $T \in \{0,1\}$ be a binary treatment variable. The quantity

$$e(x) = \mathbb{E}[T \mid X = x]$$

is known as the *propensity score* and gives the likelihood of treatment in the subpopulation defined by the condition $X = x$.

Theorem 10. *Suppose that X is admissible, and the propensity scores are positive $e(x) \neq 0$ for all X. Then,*

$$\mathbb{E}[Y \mid \mathrm{do}(T := 1)] = \mathbb{E}\left[\frac{YT}{e(X)}\right].$$

Proof. Applying the adjustment formula for a fixed y, we have

$$\begin{aligned}
\mathbb{P}[Y = y \mid \mathrm{do}(T := 1)] &= \sum_x \mathbb{P}[Y = y \mid T = 1, X = x]\,\mathbb{P}[X = x] \\
&= \sum_x \frac{\mathbb{P}[Y = y \mid T = 1, X = x]\,\mathbb{P}[X = x]\,\mathbb{P}[T = 1 \mid X = x]}{\mathbb{P}[T = 1 \mid X = x]} \\
&= \sum_x \frac{\mathbb{P}[Y = y, T = 1, X = x]}{\mathbb{P}[T = 1 \mid X = x]} \\
&= \sum_{x,t \in \{0,1\}} \frac{t\,\mathbb{P}[Y = y, T = t, X = x]}{\mathbb{P}[T = 1 \mid X = x]}.
\end{aligned}$$

Here, we used that $e(x) = \mathbb{P}[T = 1 \mid X = x] \neq 0$. Completing the proof,

$$\begin{aligned}
\mathbb{E}[Y \mid \mathrm{do}(T := 1)] &= \sum_y y\,\mathbb{P}[Y = y \mid \mathrm{do}(T := 1)] \\
&= \sum_{y,x,t \in \{0,1\}} \frac{yt\,\mathbb{P}[Y = y, T = t, X = x]}{\mathbb{P}[T = 1 \mid X = x]} = \mathbb{E}\left[\frac{YT}{e(X)}\right]. \qquad \square
\end{aligned}$$

The same theorem also shows that

$$\mathbb{E}[Y \mid \mathrm{do}(T := 0)] = \mathbb{E}\left[\frac{Y(1-T)}{1-e(X)}\right]$$

and thus the average treatment effect of X on Y is given by

$$\mathbb{E}[Y \mid \mathrm{do}(T := 1)] - \mathbb{E}[Y \mid \mathrm{do}(T := 0)] = \mathbb{E}\left[Y\left(\frac{T}{e(X)} - \frac{1-T}{1-e(X)}\right)\right].$$

This formula for the average treatment effect is called *inverse propensity score weighting*. Let's understand what it buys us compared with the adjustment formula when working with a finite sample.

One way to approximate the expectation given the theorem above is to collect many samples from which we estimate the propensity score $e(x)$ separately for each possible setting X. However, this way of going about it runs into the very same issues as the basic estimator. Practitioners therefore choose a different route.

In a first step, we fit a model $\widehat{e}$ to the propensity scores hoping that our model $\widehat{e}$ approximates the propensity score function e uniformly well. We approach this step as we would any other machine learning problem. We create a dataset of observations (x_i, e_i) where e_i is an empirical estimate of $e(x_i)$ that we compute from our sample. We then fit a model to these data points using our favorite statistical technique, be it logistic regression or something more sophisticated.

In a second step, we then use our model's estimated propensity scores in our sample estimate instead of the true propensity scores:

$$\frac{1}{n}\sum_{i=1}^{n}\frac{t_i y_i}{\widehat{e}(x_i)}.$$

The appeal of this idea is that we can use the entire repertoire of model fitting to get a good function approximation of the propensity scores. Depending on what the features are we could use logistic regression, kernel methods, random forests, or even deep models. Effectively we're reducing the problem of causal inference to that of model fitting, which we know how to do.

Double machine learning

Our previous reduction to model fitting has a notable shortcoming. The propensity score estimate $\widehat{e}(x_i)$ appears in the denominator of our estimator. This has two consequences. First, unbiased estimates of propensity scores do not imply an unbiased estimate of the causal effect. Second, when propensity scores are small and samples aren't too plentiful, this can lead to substantial variance.

There's a popular way to cope, called *double machine learning*, that works in a partially linear structural causal model:

$$Y = \tau T + g(X) + U, \qquad T = f(X) + V.$$

In this model, the variable X is an observed confounder between treatment and outcome. We allow the functions g and f to be arbitrary, but note that g only depends on X, not on T as it could in general. The random variables U, V are independent exogenous noise variables with mean 0. In this model, the effect of treatment on the outcome is linear and the coefficient τ is the desired average treatment effect.

The trick behind double machine learning is to subtract $\mathbb{E}[Y \mid X]$ from each side of the first equation and to use the fact that $\mathbb{E}[Y \mid X] = \tau\,\mathbb{E}[T \mid X] + g(X)$. We therefore get the equation

$$Y - \mathbb{E}[Y \mid X] = \tau(T - \mathbb{E}[T \mid X]) + U\,.$$

Denoting $\tilde{Y} = Y - \mathbb{E}[Y \mid X]$ and $\tilde{T} = T - \mathbb{E}[T \mid X]$ we can see that the causal effect τ is the solution to the regression problem $\tilde{Y} = \tau\tilde{T} + U$.

The idea now is to solve *two* regression problems to find good function approximations of the conditional expectations $\mathbb{E}[Y \mid X]$ and $\mathbb{E}[T \mid X]$, respectively. We can do this using data drawn from the joint distribution of (X, T, Y) by solving two subsequent model fitting problems, hence the name double machine learning.

Suppose then that we find two function approximations $q(X, Y) \approx \mathbb{E}[Y \mid X]$ and $r(X, T) \approx \mathbb{E}[T \mid X]$. We can define the random variables $\widehat{Y} = Y - q(X, Y)$ and $\widehat{T} = T - r(X, T)$. The final step is to solve the regression problem $\widehat{Y} = \widehat{\tau}\widehat{T} + U$ for the parameter $\widehat{\tau}$.

Compared with inverse propensity score weighting, we can see that finite sample errors in estimating the conditional expectations have a more benign effect on the causal effect estimate $\widehat{\tau}$. In particular, unbiased estimates of the conditional expectations lead to an unbiased estimate of the causal effect.

Heterogeneous treatment effects

In many applications, treatment effects can vary by subpopulation. In such cases we may be interested in the *conditional average treatment effect* (CATE) in the subpopulation defined by $X = x$:

$$\tau(x) = \mathbb{E}[Y \mid \mathrm{do}(T := 1), X = x] - \mathbb{E}[Y \mid \mathrm{do}(T := 0), X = x]\,.$$

We're in luck, because the same proof we saw earlier shows that we can estimate these so-called heterogeneous treatment effects with the propensity score formula:

$$\tau(x) = \mathbb{E}\left[Y\left(\frac{T}{e(X)} - \frac{1-T}{1-e(X)}\right) \mid X = x\right].$$

We can also extend double machine learning easily to the heterogeneous case by replacing the coefficient τ in the first structural equation with a function $\tau(X)$ that depends on X. The argument remains the same except that in the end we need to solve the problem $\widehat{Y} = \widehat{\tau}(X)\widehat{T} + Y$, which amounts to optimizing over a function $\widehat{\tau}$ in some model family rather than a constant $\widehat{\tau}$.

Both inverse propensity score weighting and double machine learning can, in principle, estimate heterogeneous treatment effects. These aren't the only reductions to model fitting, however. Another popular method, called *causal forests*, constructs decision trees whose leaves correspond covariate settings that deconfound treatment and outcome.[200]

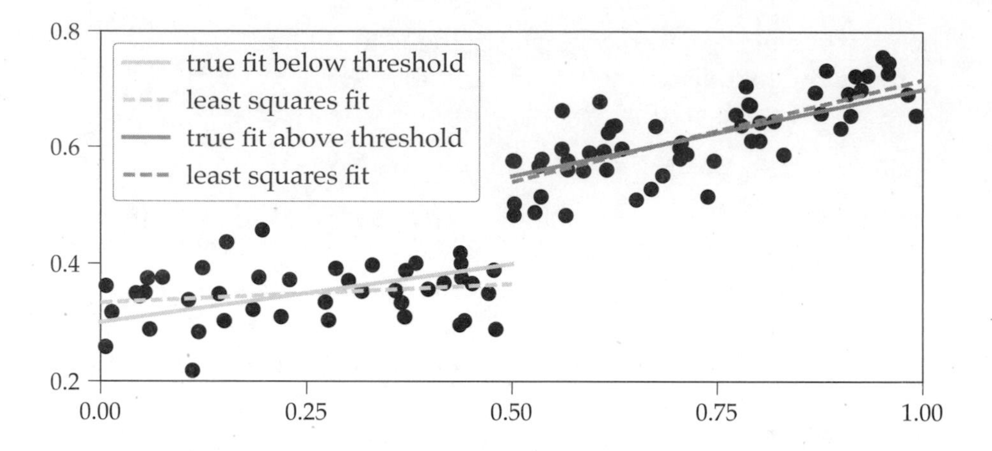

Figure 10.1: Illustration of an idealized regression discontinuity

Quasi-experiments

The idea behind quasi-experimental designs is that sometimes processes in nature or society are structured in a way that enables causal inference. The three most widely used quasi-experimental designs are *regression discontinuities*, *instrumental variables*, and *differences in differences*. We will review the first two briefly to see where machine learning comes in.

Regression discontinuity

Many consequential interventions in society trigger when a certain score R exceeds a threshold value t. The idea behind a regression discontinuity design is that units that fall just below the threshold are indistinguishable from units just above threshold. In other words, whether or not a unit is just above or just below the threshold is a matter of pure chance. We can then hope to identify a causal effect of an intervention by comparing units just below and just above the threshold.

To illustrate the idea, consider an intervention in a hospital setting that is assigned to newborn children just below a birth weight of 1,500 grams. We can ask if the intervention has a causal effect on well-being of the child at a later age as reflected in an outcome variable, such as mortality or cumulative hospital cost in their first year. We expect various factors to influence both birth weight and outcome variable. But we hope that these confounding factors are essentially held constant right around the threshold weight of 1,500 grams. Regression discontinuity designs have indeed been used to answer such questions for a number of different outcome variables.[201,202]

Once we have identified the setup for a regression discontinuity, the idea is to perform two regressions. One fits a model to the data below the threshold.

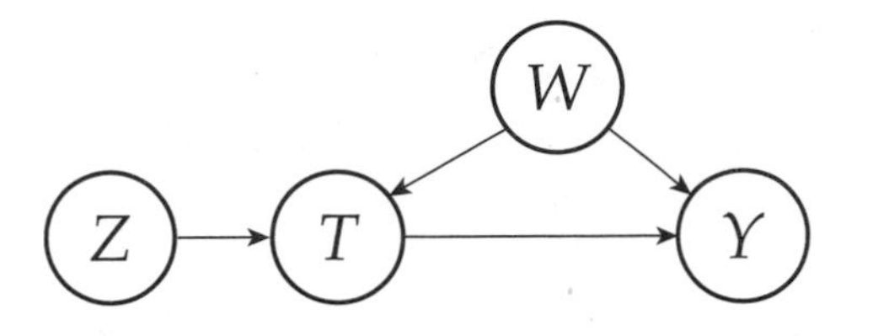

Figure 10.2: Graphical model for an instrumental variable setup

The other fits the model to data above the threshold. We then take the difference of the values that the two models predict at the threshold as our estimate of the causal effect. As usual, the idea works out nicely in an idealized linear setting and can be generalized in various ways.

There are numerous subtle and not so subtle ways a regression discontinuity design can fail. One subtle failure mode is when intervention incentivizes people to strategically make efforts to fall just below or above the threshold. Manipulation or *gaming* of the running variable is a well-known issue for instance when it comes to social program eligibility.[203] But there are other less obvious cases. For example, school class sizes in data from Chile exhibit irregularities that void regression discontinuity designs.[204] In turn, researchers have come up with tests designed to catch such problems.

Instrumental variables

Instrumental variables are a popular quasi-experimental method for causal inference. The starting point is confounding between a treatment T and our outcome of interest Y. We are in a situation where we're unable to resolve confounding via the adjustment formula. However, what we have is the existence of a special variable Z called an *instrument* that will help us estimate the treatment effect.

What makes Z a valid instrument is nicely illustrated with the following causal graph.

The graph structure encodes two key assumptions:

1. The instrument Z and the outcome Y are unconfounded.
2. The instrument Z has no direct effect on the outcome Y.

Let's walk through how this works out in the one-dimensional linear structural equation for the outcome:

$$Y = \alpha + \beta T + \gamma W + N .$$

Here, N is an independent noise term. For convenience, we denote the *error term* $U = \gamma W + N$. What we're interested in is the coefficient β since we can easily verify that it corresponds to the average treatment effect:

$$\beta = \mathbb{E}[Y \mid \mathrm{do}(T := 1)] - \mathbb{E}[Y \mid \mathrm{do}(T := 0)] \,.$$

To find the coefficient β, we cannot directly solve the regression problem $Y = \alpha + \beta T + U$, because the error term U is not independent of T due to the confounding influence of W.

However, there's a way forward after we make a few additional assumptions:

1. The error term is zero-mean: $\mathbb{E}[U] = 0$.
2. The instrument is uncorrelated with the error term: $\mathrm{Cov}(Z, U) = 0$.
3. Instrument and treatment have nonzero correlation: $\mathrm{Cov}(Z, T) \neq 0$.

The first two assumptions directly imply

$$\mathbb{E}[Y - \alpha - \beta T] = 0 \qquad \text{and} \qquad \mathbb{E}[Z(Y - \alpha - \beta T)] = 0 \,.$$

This leaves us with two linear equations in α and β so that we can solve for both parameters. Indeed, $\alpha = \mathbb{E}[Y] - \beta\,\mathbb{E}[T]$. Plugging this into the second equation, we have

$$\mathbb{E}[Z((Y - \mathbb{E}[Y]) - \beta(T - \mathbb{E}[T]))] = 0,$$

which implies, via our third assumption $\mathrm{Cov}(T, Z) \neq 0$,

$$\beta = \frac{\mathrm{Cov}(Z, Y)}{\mathrm{Cov}(T, Z)} \,.$$

There's a different intuitive way to derive this solution by solving a *two step least squares* procedure:

1. Predict the treatment from the instrument via least squares regression, resulting in the predictor $\widehat{T} = cZ$.
2. Predict the outcome from the predicted treatment using least squares regression, resulting in the predictor $\widehat{Y} = \beta' \widehat{T}$.

A calculation reveals that indeed $\beta' = \beta$, the desired treatment effect. To see this note that

$$c = \frac{\mathrm{Cov}(Z, T)}{\mathrm{Var}(Z)}$$

and hence

$$\beta' = \frac{\mathrm{Cov}(Y, \widehat{T})}{\mathrm{Var}(\widehat{T})} = \frac{\mathrm{Cov}(Y, Z)}{c\,\mathrm{Var}(Z)} = \frac{\mathrm{Cov}(Z, Y)}{\mathrm{Cov}(T, Z)} = \beta \,.$$

This solution directly generalizes to the multidimensional linear case. The two stage regression approach is in fact the way instrumental variables are often introduced operationally. We see that again instrumental variables are a clever way of reducing causal inference to prediction.

One impediment to instrumental variables is a poor correlation between the instrument and the treatment. Such instruments are called *weak instruments*. In this case, the denominator $\mathrm{Cov}(T, Z)$ in our expression for β is small and the estimation problem is ill conditioned. The other impediment is that the causal graph corresponding to instrumental variables is not necessarily easy to come by in applications. What's delicate about the graph is that we want the instrument to have a significant causal effect on the treatment, but at the same time have no other causal powers that might influence the outcome in a way that's not mediated by the treatment.

Nonetheless, researchers have found several intriguing applications of instrumental variables. One famous example goes by the name *judge instruments*. The idea is that within the United States, at least in certain jurisdictions and courts, defendants may be assigned randomly to judges. Different judges then assign different sentences, some perhaps more lenient, others harsher. The treatment here could be the sentence length and the outcome may indicate whether or not the defendant went on to commit another crime after serving the prison sentence. A perfectly random assignment of judges implies that the judge assignment and the outcome are unconfounded. Moreover, the assignment of a judge has a causal effect on the treatment, but plausibly no direct causal effect on the outcome. The assignment of judges then serves as an instrumental variable. The observation that judge assignments may be random has been the basis of much causal inference about the criminal justice system. However, the assumption of randomness in judge assignments has also been challenged.[205]

Limitations of causal inference in practice

It's worth making a distinction between causal modeling broadly speaking and the practice of causal inference today. The previous chapter covered the concepts of causal modeling. Structural causal models make it painfully clear that the model necessarily specifies strong assumptions about the data generating process. In contrast, the practice of causal inference we covered in this chapter seems almost *model-free* in how it reduces to pattern classification via technical assumptions. This appears to free the practitioner from difficult modeling choices.

The assumptions that make this all work, however, are not verifiable from data. Some papers seek assurance in statistical robustness checks, but these too are sample-based estimates. Traditional robustness checks, such as resampling methods or leave-one-out estimates, may get at issues of generalization, but cannot speak to the validity of causal assumptions.

As a result, a certain pragmatic attitude has taken hold. If we cannot verify the assumption from data anyway, we might as well make it in order to move forward. But this is a problematic position. Qualitative and theoretical ways of establishing substantive knowledge remain relevant where the limitations of data set in. The validity of a causal claim cannot be established solely based on a sample. Other sources of substantive knowledge are required.

Validity of observational methods

The empirical evidence regarding the validity of observational causal inference studies is mixed and depends on the domain of application.

A well-known article compared observational studies in the medical domain between 1985 and 1998 to the results of randomized controlled trials.[206] The conclusion was good news for observational methods:

> We found little evidence that estimates of treatment effects in observational studies reported after 1984 are either consistently larger than or qualitatively different from those obtained in randomized, controlled trials.

Another study around the same time came to a similar conclusion:

> The results of well-designed observational studies (with either a cohort or a case-control design) do not systematically overestimate the magnitude of the effects of treatment as compared with those in randomized, controlled trials on the same topic.[207]

One explanation, however, is that medical researchers may create observational designs with great care on the basis of extensive domain knowledge and prior investigation.

Freedman's paper "Statistical models and shoe leather" illustrates this point through the famous example of Jon Snow's discovery from the 1850s that cholera is a waterborne disease.[208] Many associate Snow with an early use of quantitative methods. But the application of those followed years of substantive investigation and theoretical considerations that formed the basis of the quantitative analysis.

In other domains, observational methods have been much less successful. Online advertising, for example, generates hundreds of billions of dollars in yearly global revenue, but the causal effects of targeted advertising remain a subject of debate.[209] Randomized controlled trials in this domain are rare for technical and cultural reasons. Advertising platforms are highly optimized toward a particular way of serving ads that can make true randomization difficult to implement. As a result, practitioners rely on a range of observational methods to determine the causal effect of showing an ad. However, these methods tend to perform poorly, as a recent large-scale study reveals:

> The observational methods often fail to produce the same effects as the randomized experiments, even after conditioning on extensive demographic and behavioral variables. We also characterize the incremental explanatory power our data would require to enable observational methods to successfully measure advertising effects. Our findings suggest that commonly used observational approaches based on the data usually available in the industry often fail to accurately measure the true effect of advertising.[210]

Interference, interaction, and spillovers

Confounding is not the only threat to the validity of causal studies. In a medical setting, it's often relatively easy to ensure that treatment of one subject does not influence the treatment assignment or outcome of any other unit. We called this the Stable Unit Treatment Value Assumption (SUTVA) in the previous chapter and noted that it holds by default for the units in structural causal models. Failures of SUTVA, however, are common and go by many names, such as interference, interaction, and spillover effects.

Take the example of an online social network. Interaction between units is the default in all online platforms, whose entire purpose is that people interact. Administering treatment to a subset of the platform's users typically has some influence on the control group. For example, if our treatment exposes a group of users to more content of a certain kind, those users might share the content with others outside the treatment group. In other words, treatment *spills over* to the control group. In certain cases, this problem can be mitigated by assigning treatment to a cluster in the social network that has a boundary with few outgoing edges, thus limiting bias from interaction.[211]

Interference is also common in the economic development context. To borrow an example from economist John Roemer,[212] suppose we want to know if better fishing nets would improve the yield of fishermen in a town. We design a field experiment in which we give better fishing nets to a random sample of fishermen. The results show a significantly improved yield for the treated fishermen. However, if we scale the intervention to the entire population of fishermen, we might cause overfishing and hence reduced yield for everyone.

Chapter notes

Aside from the introductory texts from the previous chapter, there are a few more particularly relevant in the context of this chapter.

The text by Angrist and Pischke[196] covers causal inference with an emphasis on regression analysis in applications in econometrics. See Athey and Imbens[213] for a more recent survey of the state of causal inference in econometrics.

Marinescu et al.[214] give a short introduction to quasi-experiments and their applications to neuroscience with a focus on regression discontinuity design, instrumental variables, and differences in differences.

Chapter 11

Sequential Decision Making and Dynamic Programming

As the previous chapters motivated, we don't just make predictions for their own sake, but rather use data to inform decision making and action. This chapter examines sequential decisions and the interplay between predictions and actions in settings where our repeated actions are directed toward a concrete end-goal. It will force us to understand statistical models that evolve over time and the nature of dependencies in data that is temporally correlated. We will also have to understand feedback and its impact on statistical decision making problems.

In machine learning, the subfield of using statistical tools to direct actions in dynamic environments is commonly called "reinforcement learning" (RL). However, this blanket term tends to lead people toward specific solution techniques. So we are going to try to maintain a broader view of the area of *sequential decision making*, including perspectives from related fields of predictive analytics and optimal control. These multiple perspectives will allow us to highlight how RL is different from the machine learning we are most familiar with.

This chapter will follow a similar flow to our study of prediction. We will formalize a temporal mathematical model of sequential decision making involving notions from *dynamical systems*. We will then present a common optimization framework for making sequential decisions when models are known: *dynamic programming*. Dynamic programming will enable algorithms for finding or approximating optimal decisions under a variety of scenarios. In the sequel, we will turn to the learning problem of how to best make sequential decisions when the mechanisms underlying dynamics and costs are not known in advance.

From predictions to actions

Let's first begin with a discussion of how sequential decision making differs from static prediction. In our study of decision theory, we laid out a framework for making optimal predictions of a binary covariate Y when we had access to data X, and probabilistic models of how X and Y were related. Supervised learning was the resulting problem of making such decisions from data rather than probabilistic models.

In sequential decision making, we add two new variables. First, we incorporate *actions*, denoted U, that we aim to take throughout a procedure. We also introduce rewards R that we aim to maximize. In sequential decision making, the goal is to analyze the data X and then subsequently choose U so that R is large. We have explicit agency in choosing U and are evaluated based on some quality scoring of U and X. There are an endless number of problems where this formulation is applied from supply chain optimization to robotic planning to online engagement maximization. Reinforcement learning is the resulting problem of taking actions so as to maximize rewards where our actions are only a function of previously observed data rather than probabilistic models. Not surprisingly, the optimization problem associated with sequential decision making is more challenging than the one that arises in decision theory.

Dynamical systems

In addition to the action variable in sequential decision making, another key feature of sequential decision making problems is the notion of time and sequence. We assume data is collected in an evolving process, and our current actions influence our future rewards.

We begin by bringing all of these elements together in the general definition of a discrete time dynamical system. The definitions are both simple and broad. We will illustrate with several examples shortly.

A *dynamical system model* has a *state* X_t, *exogenous input* U_t modeling our *control action*, and *reward* R_t. The state evolves in discrete time steps according to the equation

$$X_{t+1} = f_t(X_t, U_t, W_t)$$

where W_t is a random variable and f_t is a function. The reward is assumed to be a function of these variables as well:

$$R_t = g_t(X_t, U_t, W_t)$$

for some function g_t. To simplify our notation throughout, we will commonly write R explicitly as a function of (X, U, W), that is, $R_t[X_t, U_t, W_t]$.

Formally, we can think of this definition as a structural equation model that we also used to define causal models. After all, the equations above give us a way to incrementally build up a data generating process from noise variables. Whether or not the dynamical system is intended to capture any causal relationships in the real world is a matter of choice. Practitioners might pursue this formalism for reward maximization without modeling causal relationships. A good example is the use of sequential decision making tools for revenue maximization in targeted advertising. Rather than modeling causal relationships between, say, preferences and clicks, targeted advertising heavily relies on all sorts of signals, be they causal or not.

Concrete examples

Grocery shopping. Bob really likes to eat cheerios for breakfast every morning. Let the state X_t denotes the amount of Cheerios in Bob's kitchen on day t. The action U_t denotes the amount of Cheerios Bob buys on day t and W_t denotes the amount of Cheerios he eats that day. The random variable W_t varies with Bob's hunger. This yields the dynamical system

$$X_{t+1} = X_t + U_t - W_t\,.$$

While this example is a bit cartoonish, it turns out that such simple models are commonly used in managing large production supply chains. Any system where resources are stochastically depleted and must be replenished can be modeled comparably. If Bob had a rough model for how much he eats in a given day, he could forecast when his supply would be depleted. And he could then minimize the number of trips he'd need to make to the grocer using optimal control.

Moving objects. Consider a physical model of a flying object. The simplest dynamical model of this system is derived from Newton's laws of mechanics. Let Z_t denote the position of the vehicle (this is a three-dimensional vector). The derivative of position is velocity

$$V_t = \frac{\partial Z_t}{\partial t}\,,$$

and the derivative of velocity is acceleration

$$A_t = \frac{\partial V_t}{\partial t}\,.$$

Now, we can approximate the rules in discrete time with the simple Taylor approximations

$$\begin{aligned} Z_{t+1} &= Z_t + \Delta V_t \\ V_{t+1} &= V_t + \Delta A_t\,. \end{aligned}$$

We also know by Newton's second law that acceleration is equal to the total applied force divided by the mass of the object: $F = mA$. The flying object will be subject to external forces such as gravity and wind W_t and it will also receive forces from its propellers U_t. Then we can add this to the equations to yield a model

$$Z_{t+1} = Z_t + \Delta V_t$$
$$V_{t+1} = V_t + \frac{\Delta}{m}(W_t + U_t)\,.$$

Oftentimes, we only observe the acceleration through an accelerometer. Then estimating the position and velocity becomes a filtering problem. Optimal control problems for this model include flying the object along a given trajectory or flying to a desired location in a minimal amount of time.

Markov decision processes

Our definition in terms of structural equations is not the only dynamical model used in machine learning. Some people prefer to work directly with probabilistic transition models and conditional probabilities. In a *Markov Decision Process* (MDP), we again have a state X_t and input U_t, and they are linked by a probabilistic model

$$\mathbb{P}[X_{t+1} \mid X_t, U_t]\,.$$

This is effectively the same as the structural equation model above except we hide the randomness in this probabilistic notation.

Example: Machine repair. The following example illustrates the elegance of the conditional probability models for dynamical systems. This example comes from Bertsekas.[215] Suppose we have a machine with 10 states of repair. State 10 denotes excellent condition and 1 denotes the inability to function. Every time one uses the machine in state j, it has a probability of falling into disrepair, given by the probabilities $\mathbb{P}[X_{t+1} = i \mid X_t = j]$, where $\mathbb{P}[X_{t+1} = i \mid X_t = j] = 0$ if $i > j$. The action a one can take at any time is to repair the machine, resetting the system state to 10. Hence

$$\mathbb{P}[X_{t+1} = i \mid X_t = j, U_t = 0] = \mathbb{P}[X_{t+1} = i \mid X_t = j]$$

and

$$\mathbb{P}[X_{t+1} = i \mid X_t = j, U_t = 1] = \mathbb{1}\{i = 10\}\,.$$

While we could write this dynamical system as a structural equation model, it is more conveniently expressed by these probability tables.

Optimal sequential decision making

Just as risk minimization was the main optimization problem we studied in static decision theory, there is an abstract class of optimization problems that underlies most sequential decision making (SDM) problems. The main problem is to find a sequence of decision *policies* that maximize a cumulative reward subject to the uncertain, stochastic system dynamics. At each time, we assign a reward $R_t(X_t, U_t, W_t)$ to the current state-action pair. The goal is to find a sequence of actions to make the summed reward as large as possible:

$$\begin{array}{ll} \text{maximize}_{\{u_t\}} & \mathbb{E}_{W_t}\left[\sum_{t=0}^{T} R_t(X_t, u_t, W_t)\right] \\ \text{subject to} & X_{t+1} = f_t(X_t, u_t, W_t) \\ & (x_0 \text{ given}). \end{array}$$

Here, the expected value is over the sequence of stochastic disturbance variables W_t. Here, W_t is a random variable and X_t is hence also a random variable. The sequence of actions $\{u_t\}$ is our decision variable. It could be chosen via a random or deterministic procedure as a matter of design. But it is important to understand what information is allowed to be used in order to select u_t.

Since the dynamics are stochastic, the optimal SDM problem typically allows a policy to observe the state before deciding upon the next action. This allows a decision strategy to continually mitigate uncertainty through feedback. This is why we optimize over policies rather than over a deterministic sequence of actions. That is, our goal is to find functions of the current state π_t such that $U_t = \pi_t(X_t, X_{t-1}, \ldots)$ is optimal in expected value. By a *control policy* (or simply "a policy") we mean a function that takes a trajectory from a dynamical system and outputs a new control action. In order for π_t to be implementable, it must have access only to previous states and actions.

The policies π_t are the decision variables of the problem:

$$\begin{array}{ll} \text{maximize}_{\pi_t} & \mathbb{E}_{W_t}\left[\sum_{t=0}^{T} R_t(X_t, U_t, W_t)\right] \\ \text{subject to} & X_{t+1} = f(X_t, U_t, W_t) \\ & U_t = \pi_t(X_t, X_{t-1}, \ldots) \\ & (x_0 \text{ given}). \end{array}$$

Now, U_t is explicitly a random variable as it is a function of the state X_t.

This SDM problem will be the core of what we study in this chapter. And our study will follow a similar path to the one we took with decision theory. We will first study how to solve these SDM problems when we know the model. There is a general-purpose solution for these problems known as *dynamic programming*.

Dynamic programming

The dynamic programming solution to the SDM problem is based on the *principle of optimality*: if you've found an optimal control policy for a time horizon of length T, $\pi_1, \ldots, \pi_T$, and you want to know the optimal strategy starting at state x at time t, then you just have to take the optimal policy starting at time t, $\pi_t, \ldots, \pi_T$. The best analogy for this is based on driving directions: if you have mapped out an optimal route from Seattle to Los Angeles, and this path goes through San Francisco, then you must also have the optimal route from San Francisco to Los Angeles as the tail end of your trip. Dynamic programming is built on this principle, allowing us to recursively find an optimal policy by starting at the final time and going backwards in time to solve for the earlier stages.

To proceed, define the *Q-function* to be the mapping:

$$\begin{aligned} \mathcal{Q}_{a\to b}(x,u) &= \max_{\{u_t\}} \mathop{\mathbb{E}}_{W_t} \left[\sum_{t=a}^{b} R_t(X_t, u_t, W_t) \right] \\ \text{s.t.} \quad & X_{t+1} = f_t(X_t, u_t, W_t), \quad (X_a, u_a) = (x, u)\,. \end{aligned}$$

The Q-function determines the best achievable value of the SDM problem over times a to b when the action at time a is set to be u and the initial condition is x. It then follows that the optimal value of the SDM problem is $\max_u \mathcal{Q}_{0\to T}(x_0, u)$, and the optimal policy is $\pi(x_0) = \arg\max_u \mathcal{Q}_{0\to T}(x_0, u)$. If we had access to the Q-function for the horizon $0, T$, then we'd have everything we'd need to know to take the first step in the SDM problem. Moreover, the optimal policy is only a function of the current state of the system. Once we see the current state, we have all the information we need to predict future states, and hence we can discard the previous observations.

We can use dynamic programming to compute this Q-function and the Q-function associated with every subsequent action. That is, clearly we have that the terminal Q-function is

$$\mathcal{Q}_{T\to T}(x,u) = \mathop{\mathbb{E}}_{W_T} \left[R_T(x,u,W_T) \right] ,$$

and then compute recursively

$$\mathcal{Q}_{t\to T}(x,u) = \mathop{\mathbb{E}}_{W_t} \left[R_t(x,u,W_t) + \max_{u'} \mathcal{Q}_{t+1\to T}(f_t(x,u,W_t), u') \right] .$$

This expression is known as Bellman's equation. We also have that for all times t, the optimal policy is $u_t = \arg\max_u \mathcal{Q}_{t\to T}(x_t, u)$ and the policy depends only on the current state.

To derive this form of the Q-function, we assume inductively that this form is true for all times beyond $t+1$ and then have the chain of identities

$$\begin{aligned}
\mathcal{Q}_{t\to T}(x,u) &= \max_{\pi_{t+1},\ldots,\pi_T} \mathop{\mathbb{E}}_{w}\left[R_t(x,u,W_t) + \sum_{s=t+1}^{T} R_s(X_s,\pi_s(X_s),W_s)\right] \\
&= \mathop{\mathbb{E}}_{W_t}\left[R_t(x,u,W_t) + \max_{\pi_{t+1},\ldots,\pi_T} \mathop{\mathbb{E}}_{W_{t+1},\ldots,W_T}\left\{\sum_{s=t+1}^{T} R_s(X_s,\pi_s(X_s),W_s)\right\}\right] \\
&= \mathop{\mathbb{E}}_{W_t}\left[R_t(x,u,W_t) + \max_{\pi_{t+1}} Q\left\{f(x,u,W_t),\pi_{t+1}(f(x,u,W_t))\right\}\right] \\
&= \mathop{\mathbb{E}}_{W_t}\left[R_t(x,u,W_t) + \max_{u'} Q(f(x,u,W_t),u')\right].
\end{aligned}$$

Here, the most important point is that the maximum can be exchanged with the expectation with respect to the first W_t. This is because the policies are allowed to make decisions based on the history of observed states, and these states are deterministic functions of the noise process.

Infinite time horizons and stationary policies

The Q-functions we derived for these finite time horizons are *time varying*. One applies a different policy for each step in time. However, on long horizons with time invariant dynamics and costs, we can get a simpler formula. First, for example, consider the limit:

$$\begin{array}{ll}
\text{maximize} & \lim_{N\to\infty} \mathbb{E}_{W_t}\left[\frac{1}{N}\sum_{t=0}^{N} R(X_t,U_t,W_t)\right] \\
\text{subject to} & X_{t+1} = f(X_t,U_t,W_t),\ U_t = \pi_t(X_t) \\
& (x_0 \text{ given}).
\end{array}$$

Such infinite time horizon problems are referred to as *average cost* dynamic programs. Note that there are no subscripts on the rewards or transition functions in this model.

Average cost dynamic programming is deceptively difficult. These formulations are not directly amenable to standard dynamic programming techniques except in cases with special structure. A considerably simpler infinite time formulation is known as *discounted* dynamic programming, and this is the most popular studied formulation. Discounting is a mathematical convenience that dramatically simplifies algorithms and analysis. Consider the SDM problem

$$\begin{array}{ll}
\text{maximize} & (1-\gamma)\,\mathbb{E}_{W_t}\left[\sum_{t=0}^{\infty} \gamma^t R(X_t,U_t,W_t)\right] \\
\text{subject to} & X_{t+1} = f(X_t,U_t,W_t),\ U_t = \pi_t(X_t) \\
& (x_0 \text{ given}),
\end{array}$$

where γ is a scalar in $(0,1)$ called the *discount factor*. For γ close to 1, the discounted reward is approximately equal to the average reward. However, unlike the average cost model, the discounted cost has particularly clean optimality conditions. If we define $\mathcal{Q}_\gamma(x, u)$ to be the Q-function obtained from solving the discounted problem with initial condition x, then we have a discounted version of dynamic programming, now with the same Q-functions on the left- and right-hand sides:

$$\mathcal{Q}_\gamma(x, u) = \mathop{\mathbb{E}}_{W}\left[R(x, u, W) + \gamma \max_{u'} \mathcal{Q}_\gamma(f(x, u, W), u')\right].$$

The optimal policy is now for *all times* to let

$$u_t = \arg\max_u \mathcal{Q}_\gamma(x_t, u).$$

The policy is time invariant and one can execute it without any knowledge of the reward or dynamics functions. At every stage, one simply has to maximize a function to find the optimal action. Foreshadowing to the next chapter, the formula additionally suggests that the amount that needs to be "learned" in order to "control" is not very large for these infinite time horizon problems.

Computation

Though dynamic programming is a beautiful universal solution to the very general SDM problem, the generality also suggests computational barriers. Dynamic programming is only efficiently solvable for special cases, and we now describe a few important examples.

Tabular MDPs

Tabular MDPs refer to Markov Decision Processes with small number of states and actions. Say that there are S states and A actions. Then the the transition rules are given by tables of conditional probabilities $\mathbb{P}[X_{t+1}|X_t, U_t]$, and the size of such tables are S^2A. The Q-functions for the tabular case are also tables, each of size SA, enumerating the cost-to-go for all possible state action pairs. In this case, the maximization

$$\max_{u'} \mathcal{Q}_{a\to b}(x, u')$$

corresponds to looking through all of the actions and choosing the largest entry in the table. Hence, in the case that the rewards are deterministic functions of x and u, Bellman's equation simplifies to

$$\mathcal{Q}_{t\to T}(x, u) = R_t(x, u) + \sum_{x'} \mathbb{P}[X_{t+1} = x'|X_t = x, U_t = u] \max_{u'} \mathcal{Q}_{t+1\to T}(x', u').$$

This function can be computed by elementary matrix-vector operations: the Q-functions are $S \times A$ arrays of numbers. The "max" operation can be performed by operating over each row in such an array. The summation with respect to x' can be implemented by multiplying a $SA \times S$ array by an S-dimensional vector. We complete the calculation by summing the resulting expression with the $S \times A$ array of rewards. Hence, the total time to compute $\mathcal{Q}_{t \to T}$ is $O(S^2 A)$.

Linear quadratic regulator

The other important problem where dynamic programming is efficiently solvable is the case when the dynamics are *linear* and the rewards are *quadratic*. In control design, this class of problems is generally referred to as the problem of the Linear Quadratic Regulator (LQR):

$$\begin{array}{ll} \text{minimize} & \mathbb{E}_{W_t}\left[\frac{1}{2}\sum_{t=0}^{T} X_t^T \Phi_t X_t + U_t^T \Psi_t U_t\right], \\ \text{subject to} & X_{t+1} = A_t X_t + B_t U_t + W_t,\ U_t = \pi_t(X_t) \\ & (x_0 \text{ given}). \end{array}$$

Here, Φ_t and Ψ_t are most commonly positive semidefinite matrices. w_t is noise with zero mean and bounded variance, and we assume W_t and $W_{t'}$ are independent when $t \neq t'$. The state transitions are governed by a linear update rule with A_t and B_t appropriately sized matrices. We also abide by the common convention in control textbooks to pose the problem as a minimization—not maximization—problem.

As we have seen above, many systems can be modeled by linear dynamics in the real world. However, we haven't yet discussed cost functions. It's important to emphasize here that cost functions are *designed* not given. Recall back to supervised learning: though we wanted to minimize the number of errors made on out-of-sample data, on in-sample data we minimized convex surrogate problems. The situation is exactly the same in this more complex world of dynamical decision making. Cost functions are designed by the engineer so that the SDM problems are tractable but also so that the desired outcomes are achieved. Cost function design is part of the toolkit for online decision making, and quadratic costs can often yield surprisingly good performance for complex problems.

Quadratic costs are also attractive for computational reasons. They are convex as long as Φ_t and Ψ_t are positive definite. Quadratic functions are closed under minimization, maximization, addition. And for zero mean noise W_t with covariance Σ, we know that the noise interacts nicely with the cost function. That is, we have

$$\mathbb{E}_W[(x+W)^T M(x+W)] = x^T M x + \mathrm{Tr}(Q\Sigma)$$

for any vector x and matrix M. Hence, when we run dynamic programming, every Q-function is necessarily quadratic. Moreover, since the Q-functions are quadratic, the optimal action is a *linear function* of the state

$$U_t = -K_t X_t$$

for some matrix K_t.

Now consider the case where there are static costs $\Phi_t = \Phi$ and $\Psi_t = \Psi$, and time invariant dynamics such that $A_t = A$ and $B_t = B$ for all t. One can check that the Q-function on a finite time horizon satisfies a recursion

$$\mathcal{Q}_{t \to T}(x, u) = x^T \Phi x + u^T \Psi u + (Ax + Bu)^T M_{t+1}(Ax + Bu) + c_t \, .$$

for some positive definite matrix M_{t+1}. In the limit as the time horizon tends to infinity, the optimal control *policy* is *static, linear state feedback*:

$$u_t = -K x_t \, .$$

Here the matrix K is defined by

$$K = (\Psi + B^T M B)^{-1} B^T M A$$

and M is a solution to the *Discrete Algebraic Riccati Equation*

$$M = \Phi + A^T M A - (A^T M B)(\Psi + B^T M B)^{-1}(B^T M A) \, .$$

Here, M is the unique solution of the Riccati equation where all of the eigenvalues of $A - BK$ have magnitude less than 1. Finding this specific solution is relatively easy using standard linear algebraic techniques. It is also the limit of the Q-functions computed above.

Policy and value iteration

Two of the most well studied methods for solving such discounted infinite time horizon problems are *value iteration* and *policy iteration*. Value iteration proceeds by the steps

$$\mathcal{Q}_{k+1}(x, u) = \mathop{\mathbb{E}}_{W} \left[R(x, u, W) + \gamma \max_{u'} \mathcal{Q}_k(f(x, u, W), u') \right] \, .$$

That is, it simply tries to solve the Bellman equation by running a fixed point operation. This method succeeds when the iteration is a contraction mapping, and this occurs in many contexts.

On the other hand, policy iteration is a two step procedure: *policy evaluation* followed by *policy improvement*. Given a policy π_k, the policy evaluation step is given by

$$\mathcal{Q}_{k+1}(x, u) = \mathbb{E}\left[R(x, u, W) + \gamma \mathcal{Q}_k(f(x, \pi_k(x), W), \pi_k(x))\right] \, .$$

And then the policy is updated by the rule

$$\pi_{k+1}(x) = \arg\max_u \mathcal{Q}_{k+1}(x,u)\,.$$

Oftentimes, several steps of policy evaluation are performed before updating the policy.

For both policy and value iteration, we need to be able to compute expectations efficiently and must be able to update *all values* of x and u in the associated Q functions. This is certainly doable for tabular MDPs. For low-dimensional problems, policy iteration and value iteration can be approximated by gridding state space, and then treating the problem as a tabular one. Then, the resulting Q function can be extended to other (x,u) pairs by interpolation. There are also special cases where the maxima and minima yield closed form solutions and hence these iterations reduce to simpler forms. LQR is a canonical example of such a situation.

Model predictive control

If the Q-functions in value or policy iteration converge quickly, long-term planning might not be necessary, and we can effectively solve infinite horizon problem with short-term planning. This is the key idea behind one of the most powerful techniques for efficiently and effectively finding quality policies for SDM problems called *model predictive control.*

Suppose that we aim to solve the infinite horizon average reward problem:

$$\begin{array}{ll} \text{maximize} & \lim_{T\to\infty} \mathbb{E}_{W_t}\left[\frac{1}{T}\sum_{t=0}^{T} R_t(W_t,U_t)\right] \\ \text{subject to} & X_{t+1} = f_t(X_t,U_t,W_t) \\ & U_t = \pi_t(X_t) \\ & (x_0 \text{ given}). \end{array}$$

Model Predictive Control computes an *open loop* policy on a finite horizon H

$$\begin{array}{ll} \text{maximize}_{u_t} & \mathbb{E}_{W_t}\left[\sum_{t=0}^{H} R_t(X_t,u_t)\right] \\ \text{subject to} & X_{t+1} = f_t(X_t,u_t,W_t) \\ & (X_0 = \text{x}). \end{array}$$

This gives a sequence $u_0(x),\ldots,u_H(x)$. The policy is then set to be $\pi(x) = u_0(x)$. After this policy is executed, we observe a new state, x', based on the dynamics. We then recompute the optimization, now using $x_0 = x'$ and setting the action to be $\pi(x') = u_0(x')$.

MPC is a rather intuitive decision strategy. The main idea is to plan out a sequence of actions for a given horizon, taking into account as much uncertainty as possible. But rather than executing the entire sequence, we play the first action and then gain information from the environment about the noise. This direct feedback influences the next planning stage. For this reason, model predictive control is often a successful control policy even when implemented with inaccurate or approximate models. Model Predictive Control also allows us to easily add a variety of constraints to our plan at little cost, such as bounds on the magnitude of the actions. We just append these to the optimization formulation and then lean on the computational solver to make us a plan with these constraints.

To concretely see how Model Predictive Control can be effective, it's helpful to work through an example. Let's suppose the dynamics and rewards are time invariant. Let's suppose further that the reward function is bounded above, and there is some state-action pair $(x_\star, u_\star)$ which achieves this maximal reward R_{max}.

Suppose we solve the finite time horizon problem where we enforce that (x, u) must be at $(x_\star, u_\star)$ at the end of the time horizon:

$$\begin{array}{ll} \text{maximize} & \mathbb{E}_{W_t}[\sum_{t=0}^{H} R(X_t, u_t)] \\ \text{subject to} & X_{t+1} = f(X_t, u_t, W_t) \\ & (X_H, U_H) = (x_\star, u_\star) \\ & (X_0 = \mathrm{x}). \end{array}$$

We replan at every time step by solving this optimization problem and taking the first action.

The following proposition summarizes how this policy performs

Proposition 11. *Assume that all rewards are bounded above by R_{max}. Then with the above MPC policy, we have for all T that*

$$\mathbb{E}\left[\frac{1}{T}\sum_{t=0}^{T} R(x_t, u_t)\right] \geq \frac{\mathcal{Q}_{0\to H}(x_0, u_0) - HR_{\text{max}}}{T} + \mathbb{E}_W[R(f(x_\star, u_\star, W), 0)]\,.$$

The proposition asserts that there is a "burn in" cost associated with the initial horizon length. This term goes to zero with T, but will have different values for different H. The policy converges to a residual average cost due to the stochasticity of the problem and the fact that we try to force the system to the state $(x_\star, u_\star)$.

Proof. To analyze how the policy performs, we turn to Bellman's equation. For any time t, the MPC policy is

$$u_t = \arg\max_u \mathcal{Q}_{0\to H}(x_t, u)$$

Now,

$$\mathcal{Q}_{0\to H}(x_t, u) = R(x_t, u) + \mathbb{E}[\max_{u'} \mathcal{Q}_{1\to H}(X_{t+1}, u')]\,.$$

Now consider what to do at the time $t+1$. A *suboptimal* strategy at this time is to try to play the optimal strategy on the horizon $1 \to H$, and then do nothing on the last step. That is,

$$\max_u \mathcal{Q}_{0\to H}(x_{t+1}, u) \geq \max_{u'} \mathcal{Q}_{1\to H}(x_{t+1}, u') + \mathbb{E}[R(f(x_\star, u_\star, W_{t+H}), 0)] .$$

The last expression follows because the action sequence from $1 \to H$ enforces $(x_{t+H}, u_{t+H}) = (x_\star, u_\star)$. The first term on the right hand side was computed in expectation above, hence we have

$$\mathbb{E}[\max_u \mathcal{Q}_{0\to H}(X_{t+1}, u)] \geq \mathbb{E}[\mathcal{Q}_{0\to H}(x_t, u_t)] - \mathbb{E}[R(x_t, u_t, W)] \\ + \mathbb{E}[R(f(x_\star, u_\star, W), 0)] .$$

Unwinding this recursion, we find

$$\mathbb{E}[\max_u \mathcal{Q}_{0\to H}(X_{T+1}, u)] \geq \mathcal{Q}_{0\to H}(x_0, u_0) - \mathbb{E}\left[\sum_{t=0}^{T} R(x_t, u_t, W_t)\right] \\ + T\mathbb{E}[R(f(x_\star, u_\star, W), 0)] .$$

Since the rewards are bounded above, we can upper bound the left hand side by $R_{\max}H$. Rearranging terms then proves the theorem.

□

The main caveat with this argument is that there may not *exist* a policy that drives x to $x_\star$ from an arbitrary initial condition and any realization of the disturbance signal. Much of the analysis of MPC schemes is devoted to guaranteeing that the problems are *recursively feasible*, meaning that such constraints can be met for all time.

This example also shows how it is often helpful to have some sort of recourse at the end of the planning horizon to mitigate the possibility of being too greedy and driving the system into a bad state. The terminal condition of forcing $x_H = 0$ adds an element of safety to the planning, and ensures stable execution for all time. More general, adding some terminal condition to the planning horizon $C(x_H)$ is part of good Model Predictive Control design and is often a powerful way to balance performance and robustness.

Partial observation and the separation heuristic

Let's now move to the situation where instead of observing the state directly, we observe an output Y_t:

$$Y_t = h_t(X_t, U_t, W_t) .$$

All of the policies we derived from optimization formulations above required feeding back a function of the state. When we can only act on outputs, SDM problems are considerably more difficult.

1. **Static Output Feedback is NP-hard.** Consider the case of just building a static policy from the output Y_t. Let's suppose our model is just the simple linear model:

$$X_{t+1} = AX_t + BU_t$$
$$Y_t = CX_t\,.$$

   Here, A, B and C are matrices. Suppose we want to find a feedback policy $U_t = KY_t$ (where K is a matrix) and all we want to guarantee is that for any initial x_0, the system state converges to zero. This problem is called *static state feedback* and is surprisingly NP-hard. It turns out that the problem is equivalent to finding a matrix K such that $A + BKC$ has all of its eigenvalues inside the unit circle in the complex plane.[216] In the MDP case, static state feedback was not only optimal but also computable for tabular MDPs and certain other SDM problems. By contrast, static output feedback is computationally intractable.

2. **POMDPs are PSPACE hard.** Papadimitriou and Tsitsiklis showed that optimization of POMDPs—i.e., MDPs where some states are partially observed—is intractable even on small state spaces.[217] They reduced the problem of *quantifier elimination* in logical satisfiability problems (QSAT) to POMDPs. QSAT seeks to determine the validity of statements like "there exists x such that for all y there exists z such that for all w this logical formula is true." Optimal actions in POMDPs essentially have to keep track of all of the possible true states that might have been visited given the partial observation and make actions accordingly. Hence, the policies have a similar flavor to quantifier elimination as they seek actions that are beneficial to all possible occurrences of the unobserved variables. Since these policies act over long time horizons, the number of counterfactuals that must be maintained grows exponentially large.

Despite these challenges, engineers solve POMDP problems all the time. Just because the problems are hard in general doesn't mean they are intractable on average. It only means that we cannot expect to have general-purpose optimal algorithms for these problems. Fortunately, suboptimal solutions are oftentimes quite good for practice, and there are many useful heuristics for decision making with partial information. The most common approach to the output feedback problem is the following two-stage strategy:

1. **Filtering.** Using all of your past data $\{y_s\}$ for $s = 0, \dots, t$, build an estimate, $\widehat{x}_t$, of your state.

2. **Action based on certainty equivalence.** Solve the desired SDM problem as if you had perfect observation of the state X_t, using $\widehat{x}_t$ wherever you would have used an observation $X_t = x_t$. At runtime, plug in the estimator $\widehat{x}_t$ as if it were a perfect measurement of the state.

This strategy uses a *separation principle* between prediction and action. For certain problems, this two-staged approach is actually optimal. Notably, if the SDM problem has quadratic rewards/costs, if the dynamics are linear, and if the noise process is Gaussian, then the separation between prediction and action is optimal. More commonly, the separation heuristic is suboptimal, but this abstraction also enables a simple heuristic that is easy to debug and simple to design.

While we have already covered algorithms for optimal control, we have not yet discussed state estimation. Estimating the state of a dynamical system receives the special name *filtering*. However, at the end of the day, filtering is a prediction problem. Define the observed data up to time t as

$$\tau_t := (y_t, \ldots, y_1, u_{t-1}, \ldots, u_1) .$$

The goal of filtering is to estimate a function $h(\tau_t)$ that predicts X_t. We now describe two approaches to filtering.

Optimal filtering

Given a model of the state transition function and the observation function, we can attempt to compute the maximum a posteriori estimate of the state from the data. That is, we could compute $p(x_t|\tau_t)$ and then estimate the mode of this density. Here, we show that such an estimator has a relatively simple recursive formula, though it is not always computationally tractable to compute this formula.

To proceed, we first need a calculation that takes advantage of the conditional independence structure of our dynamical system model. Note that

$$\begin{aligned}
p(y_t, x_t, x_{t-1}, u_{t-1}|\tau_{t-1}) &= \\
&p(y_t|x_t, x_{t-1}, u_{t-1}, \tau_{t-1})p(x_t|x_{t-1}, u_{t-1}, \tau_{t-1}) \\
&\qquad \times p(x_{t-1}|\tau_{t-1})p(u_{t-1}|\tau_{t-1}) \\
&= p(y_t|x_t)p(x_t|x_{t-1}, u_{t-1})p(x_{t-1}|\tau_{t-1})p(u_{t-1}|\tau_{t-1}) .
\end{aligned}$$

This decomposition is into terms we now recognize. The terms $p(x_t|x_{t-1}, u_{t-1})$ and $p(y_t|x_t)$ define the POMDP model and are known. The policy $p(u_t|\tau_t)$ is what we're trying to design. The only unknown here is $p(x_{t-1}|\tau_{t-1})$, but this expression gives us a recursive formula to $p(x_t|\tau_t)$ for all t.

To derive this formula, apply Bayes' rule and then use the above calculation:

$$
\begin{aligned}
p(x_t|\tau_t) &= \frac{\int_{x_{t-1}} p(x_t, y_t, x_{t-1}, u_{t-1}|\tau_{t-1})}{\int_{x_t, x_{t-1}} p(x_t, y_t, x_{t-1}, u_{t-1}|\tau_{t-1})} \\
&= \frac{\int_{x_{t-1}} p(y_t|x_t)p(x_t|x_{t-1}, u_{t-1})p(x_{t-1}|\tau_{t-1})p(u_{t-1}|\tau_{t-1})}{\int_{x_t, x_{t-1}} p(y_t|x_t)p(x_t|x_{t-1}, u_{t-1})p(x_{t-1}|\tau_{t-1})p(u_{t-1}|\tau_{t-1})} \\
&= \frac{\int_{x_{t-1}} p(y_t|x_t)p(x_t|x_{t-1}, u_{t-1})p(x_{t-1}|\tau_{t-1})}{\int_{x_t, x_{t-1}} p(y_t|x_t)p(x_t|x_{t-1}, u_{t-1})p(x_{t-1}|\tau_{t-1})} \,. \qquad (4)
\end{aligned}
$$

Given a prior for x_0, this now gives us a formula to compute a MAP estimate of x_t for all t, incorporating data in a streaming fashion. For tabular POMDP models with small state spaces, this formula can be computed simply by summing up the conditional probabilities. In POMDPs without inputs—also known as *hidden Markov models*—this formula gives the forward pass of Viterbi's decoding algorithm. For models where the dynamics are linear and the noise is Gaussian, these formulas reduce into an elegant closed form solution known as *Kalman filtering*. In general, this optimal filtering algorithm is called *belief propagation* and is the basis of a variety of algorithmic techniques in the field of graphical models.

Kalman filtering

For the case of linear dynamical systems, the above calculation has a simple closed form solution that looks similar to the solution of LQR. This estimator is called a *Kalman filter*, and is one of the most important tools in signal processing and estimation. The Kalman filter assumes a linear dynamical system driven by Gaussian noise with observations corrupted by Gaussian noise

$$
\begin{aligned}
X_{t+1} &= AX_t + BU_t + W_t \\
Y_t &= CX_t + V_t \,.
\end{aligned}
$$

Here, assume W_t and V_t are independent for all time and Gaussian with means zero and covariances Σ_W and Σ_V respectively. Because of the joint Gaussianity, we can compute a closed form formula for the density of X_t conditioned on the past observations and actions, $p(x_t|\tau_t)$. Indeed, X_t is a Gaussian itself.

On an infinite time horizon, the Kalman filter takes a simple and elucidating form:

$$
\begin{aligned}
\widehat{x}_{t+1} &= A\widehat{x}_t + Bu_t - L(y_t - \widehat{y}_t) \\
\widehat{y}_t &= C\widehat{x}_t
\end{aligned}
$$

where

$$
L = APC^T(CPC^T + \Sigma_V)^{-1}
$$

and P is the positive semidefinite solution to the discrete algebraic Riccati equation

$$P = APA^T + \Sigma_W - (APC^T)(CPC^T + \Sigma_V)^{-1}(C\Sigma A^T).$$

The derivation of this form follows from the calculation in Equation 4. But the explanation of the formulas tends to be more insightful than the derivation. Imagine the case where $L = 0$. Then our estimate $\widehat{x}_t$ simulates the same dynamics as our model with no noise corruption. The matrix L computes a correction for this simulation based on the observed y_t. This feedback correction is chosen in such a way such that P is the steady state covariance of the error $X_t - \widehat{x}_t$. P ends up being the minimum variance possible with an estimator that is *unbiased* in the sense that $\mathbb{E}[\widehat{x}_t - X_t] = 0$.

Another interesting property of this calculation is that the L matrix is the LQR gain associated with the LQR problem

$$\begin{array}{ll} \text{minimize} & \lim_{T\to\infty} \mathbb{E}_{W_t}\left[\frac{1}{2}\sum_{t=0}^{T} X_t^T \Sigma_W X_t + U_t^T \Sigma_V U_t\right], \\ \text{subject to} & X_{t+1} = A^T X_t + B^T U_t + W_t,\ U_t = \pi_t(X_t) \\ & (x_0 \text{ given}). \end{array}$$

Control theorists often refer to this pairing as the *duality between estimation and control*.

Feedforward prediction

While the optimal filter can be computed in simple cases, we often do not have simple computational means to compute the optimal state estimate. That said, the problem of state estimation is necessarily one of prediction, and the first half of this course gave us a general strategy for building such estimators from data. Given many simulations or experimental measurements of our system, we can try to estimate a function h such that $X_t \approx h(\tau_t)$. To make this concrete, we can look at *time lags* of the history,

$$\tau_{t-s\to t} := (y_t, \ldots, y_{t-s}, u_{t-1}, \ldots, u_{t-s}).$$

Such time lags are necessarily all of the same length. Then estimating

$$\text{minimize}_h \quad \textstyle\sum_t \text{loss}(h(\tau_{t-s\to t}), x_t)$$

is a supervised learning problem, and standard tools can be applied to design architectures for and estimate h.

Chapter notes

This chapter and the following chapter overlap significantly with a survey of reinforcement learning by Recht,[218] which contains additional connections to continuous control. Those interested in learning more about continuous control from an optimization viewpoint should consult the book by Borrelli et al.[219] This book also provides an excellent introduction to model predictive control. Another excellent introduction to continuous optimal control and filtering is Boyd's lecture notes.[220]

An invaluable introduction to the subject of dynamic programming is by Bertsekas, who has done pioneering research in this space and has written some of the most widely read texts.[215] For readers interested in a mathematical introduction to dynamic programming on discrete processes, we recommend Puterman's text.[221] Puterman also explains the linear programming formulation of dynamic programming.

Chapter 12

Reinforcement Learning

Dynamic programming and its approximations studied thus far all require knowledge of the probabilistic mechanisms underlying how data and rewards change over time. When these mechanisms are unknown, appropriate techniques to probe and learn about the underlying dynamics must be employed in order to find optimal actions. We shall refer to the solutions to sequential decision making problems when the dynamics are unknown as *reinforcement learning*. Depending on context, the term may refer to a body of work in artificial intelligence, the community of researchers and practitioners who apply a certain set of tools to sequential decision making, and data-driven dynamic programming. That said, it is a useful name to collect a set of problems of broad interest to machine learning, control theory, and robotics communities.

A particularly simple and effective strategy for reinforcement learning problems is to estimate a predictive model for the dynamical system and then to use the fit model as if it were the true model in the optimal control problem. This is an application of the *principle of certainty equivalence*, an idea tracing back to the dynamic programming literature of the 1950s.[222,223] Certainty equivalence is a general solution strategy for the following problem. Suppose you want to solve some optimization problem with a parameter ϑ that is unknown. However, suppose we can gather data to estimate ϑ. Then the certainty equivalent solution is to use a point estimate for ϑ as if it were the true value. That is, you act as if you were certain of the value of ϑ, even though you have only estimated ϑ from data. We will see throughout this chapter that such certainty equivalent solutions are powerfully simple and effective baselines for sequential decision making in the absence of well-specified models.

Certainty equivalence is a very general principle. We can apply it to the output of a filtering scheme that predicts state, as we described in our discussion of partially observed Markov Decision Processes. We can also apply this principle in the study of MDPs with unknown parameters. For every problem in this chapter, our core baseline will always be the certainty equivalent solution. Surprisingly, we will see that certainty equivalent baselines are typically quite competitive and give a clear illustration of the best quality one can expect in many reinforcement learning problems.

Exploration-exploitation trade-offs: Regret and PAC-error

In order to compare different approaches to reinforcement learning, we need to decide on some appropriate rules of comparison. Though there are a variety of important metrics for engineering practice that must be considered, including ease of implementation and robustness, a first-order statistical comparison might ask how many samples are needed to achieve a policy with high reward.

For this question, there are two predominant conventions to compare methods: *PAC-error* and *regret*. PAC is a shorthand for *probably approximately correct*. It is a useful notion when we spend all of our time learning about a system, and then want to know how suboptimal our solution will be when build from the data gathered thus far. Regret is more geared toward online execution where we evaluate the reward accrued at all time steps, even if we are spending that time probing the system to learn about its dynamics. Our focus in this chapter will be showing that these two concepts are closely related.

Let us formalize the two notions. As in the previous chapter, we will be concerned with sequential decision making problems of the form

$$\begin{array}{ll} \text{maximize}_{\pi_t} & \mathbb{E}_{W_t}\left[\sum_{t=0}^{T} R_t(X_t, U_t, W_t)\right] \\ \text{subject to} & X_{t+1} = f(X_t, U_t, W_t) \\ & U_t = \pi_t(X_t, X_{t-1}, \dots) \\ & (x_0 \text{ given}) \,. \end{array}$$

Let $\pi_\star$ denote the optimal policy of this problem.

For PAC, let's suppose we allocate N samples to probe the system and use them in some way to build a policy π_N. We can define the optimization error of this policy to be

$$\mathcal{E}(\pi_N) = \mathbb{E}\left[\sum_{t=1}^{T} R_t[X_t', \pi_\star(X_t'), W_t]\right] - \mathbb{E}\left[\sum_{t=1}^{T} R_t[X_t, \pi_N(X_t), W_t]\right] .$$

Our model has (δ, ϵ)-PAC-error if $\mathcal{E}(\pi_N) \leq \epsilon$ with probability at least $1 - \delta$. The probability here is measured with respect to the sampling process and dynamics of the system.

Regret is defined similarly, but is subtly different. Suppose we are now only allowed T total actions and we want to understand the cumulative award achieved after applying these T actions. In this case, we have to balance the number of inputs we use to find a good policy (exploration) against the number of inputs used to achieve the best reward (exploitation).

Formally, suppose we use a policy π_t at each time step to choose our action. Suppose $\pi_\star$ is some other fixed policy. Let X_t denote the states induced by the policy sequence π_t and X_t' denote the states induced by $\pi_\star$. Then the *regret* of $\{\pi_t\}$ is defined to be

$$\mathcal{R}_T(\{\pi_t\}) = \mathbb{E}\left[\sum_{t=1}^{T} R_t[X_t', \pi_\star(X_t'), W_t]\right] - \mathbb{E}\left[\sum_{t=1}^{T} R_t[X_t, \pi_t(X_t), W_t]\right] .$$

It is simply the expected difference in the rewards generated under policy $\pi_\star$ as compared to those generated under policy sequence π_t. One major way that regret differs from PAC-error is that the policy can change with each time step.

One note of caution for both of these metrics is that they are comparing to a policy $\pi_\star$. It's possible that the comparison policy $\pi_\star$ is not very good. So we can have small regret and still not have a particularly useful solution to the SDM problem. As a designer it's imperative to understand $\pi_\star$ to formalize the best possible outcome with perfect information. That said, regret and PAC-error are valuable ways to quantify how much exploration is necessary to find nearly optimal policies. Moreover, both notions have provided successful frameworks for algorithm development: many algorithms with low regret or PAC-error are indeed powerful in practice.

Multi-armed bandits

The multi-armed bandit is one of the simplest reinforcement learning problems, and studying this particular problem provides many insights into exploration-exploitation trade-offs.

In the multi-armed bandit, we assume *no state whatsoever*. There are K total actions, and the reward is a random function of which action you choose. We can model this by saying there are i.i.d. random variables $W_{t1}, \ldots, W_{tk}$, and your reward is the dot product

$$R_t = [W_{t1}, \ldots, W_{tK}] e_{u_t}$$

where e_i is a standard basis vector. Here W_{ti} take values in the range $[0, 1]$. We assume that all of the W_{ti} are independent, and that W_{ti} and W_{si} are identically distributed. Let $\mu_i = \mathbb{E}[W_{ti}]$. Then the expected reward at time t is precisely μ_{u_t}.

The multi-armed bandit problem is inspired by gambling on slot machines. Indeed, a "bandit" is a colloquial name for a slot machine. Assume that you have K slot machines. Each machine has some probability of paying out when you play it. You want to find the machine that has the largest probability of paying out, and then play that machine for the rest of time. The reader should ponder the irony that much of our understanding of statistical decision making comes from gambling.

First let's understand what the optimal policy is if we know the model. The total reward is equal to

$$\mathop{\mathbb{E}}_{W_t}\left[\sum_{t=0}^{T} R_t(u_t, W_t)\right] = \sum_{t=1}^{T} \mu_{u_t}\,.$$

The optimal policy is hence to choose a constant action $u_t = k$ where $k = \arg\max_i \mu_i$.

When we don't know the model, it makes sense that our goal is to quickly find the action corresponding to the largest mean. Let's first do a simple PAC analysis, and then turn to the slightly more complicated regret analysis. Our simple baseline is one of certainty equivalence. We will try each action N/K times, and compute the empirical return. The empirical means are:

$$\widehat{\mu}_k = \frac{K}{N}\sum_{i=1}^{N/K} R_i^{(k)}\,.$$

Our policy will be to take the action with the highest observed empirical return.

To estimate the value of this policy, let's assume that the best action is $u = 1$. Then define

$$\Delta_i = \mu_1 - \mu_i\,.$$

Then we have

$$\mathcal{E}(\pi_N) = \sum_{i=1}^{K} T\Delta_i\,\mathbb{P}[\forall i\colon \widehat{\mu}_k \geq \widehat{\mu}_i]\,.$$

We can bound the probability that action k is selected as follows. First, if action k has the largest empirical mean, it must have a larger empirical mean than the true best option, action 1:

$$\mathbb{P}[\forall i\colon \widehat{\mu}_k \geq \widehat{\mu}_i] \leq \mathbb{P}[\widehat{\mu}_k \geq \widehat{\mu}_1]\,.$$

We can bound this last term using Hoeffding's inequality. Let $m = N/K$. Since each reward corresponds to an independent draw of some random process, we have $\widehat{\mu}_k - \widehat{\mu}_1$ is the mean of $2m$ independent random variables in the range $[-1, 1]$:

$$\frac{1}{2}(\widehat{\mu}_k - \widehat{\mu}_1) = \frac{1}{2m}\left(\sum_{i=1}^{m} R_i^{(k)} + \sum_{i=1}^{m} -R_i^{(1)}\right)\,.$$

Now writing Hoeffding's inequality for this random variable gives the tail bound

$$\mathbb{P}[\widehat{\mu}_k \geq \widehat{\mu}_1] = \mathbb{P}[\tfrac{1}{2}(\widehat{\mu}_k - \widehat{\mu}_1) \geq 0] \leq \exp\left(-\frac{m\Delta_k^2}{4}\right)\,,$$

which results in an optimization error

$$\mathcal{E}(\pi_N) \leq \sum_{i=1}^{K} T\Delta_i \exp\left(-\frac{N\Delta_i^2}{4K}\right)$$

with probability 1. This expression reveals that the multi-armed bandit problem is fairly simple. If Δ_i are all small, then any action will yield about the same reward. But if all of the Δ_i are large, then finding the optimal action only takes a few samples. Naively, without knowing anything about the gaps at all, we can use the fact that $xe^{-x^2/2} \leq \frac{1}{2}$ for nonnegative x to find

$$\mathcal{E}(\pi_N) \leq \frac{K^{3/2}T}{\sqrt{N}}\,.$$

This shows that no matter what the gaps are, as long as N is larger than K^3, we would expect to have a high-quality solution.

Let's now turn to analyzing regret of a simple certainty equivalence baseline. Given a time horizon T, we can spend the first m time steps searching for the best return. Then we can choose this action for the remaining $T - m$ time steps. This strategy is called *explore-then-commit*.

The analysis of the explore-then-commit strategy for the multi-armed bandit is a straightforward extension of the PAC analysis. If at round t we apply action k, the expected gap between our policy and the optimal policy is Δ_k. So if we let T_k denote the number of times action k is chosen by our policy then we must have

$$\mathcal{R}_T = \sum_{k=1}^{K} \mathbb{E}[T_k]\Delta_k\,.$$

T_k are necessarily random variables: what the policy learns about the different means will depend on the observed sequence x_k, which are all random variables.

Suppose that for exploration we mimic our offline procedure, trying each action m times, and recording the observed rewards for that action $r_i^{(k)}$ for $i = 1, \ldots, m$. At the end of these mk actions, we compute the empirical mean associated with each action as before. Then we must have that

$$\mathbb{E}[T_k] = m + (T - mK)\,\mathbb{P}[\forall i\colon \widehat{\mu}_k \geq \widehat{\mu}_i]\,.$$

The first term just states that each action is performed m times. The second term states that action k is chosen for the commit phase only if its empirical mean is larger than all of the other empirical means.

Using Hoeffding's inequality again to bound these probabilities, we can put everything together to bound the expected regret as

$$\mathcal{R}_T \leq \sum_{k=1}^{K} m\Delta_k + (T - mK)\Delta_k \exp\left(-\frac{m\Delta_k^2}{4}\right)\,.$$

Let's specialize to the case of *two* actions to see what we can take away from this decomposition:

1. **Gap-dependent regret.** First, assume we know the gap between the means, Δ_2, but we don't know which action leads to the higher mean. Suppose that
$$m_0 = \left\lceil \frac{4}{\Delta_2^2} \log\left(\frac{T\Delta_2^2}{4}\right) \right\rceil \geq 1 .$$
Then using $m = m_0$, we have
$$\begin{aligned} \mathcal{R}_T &\leq m\Delta_2 + T\Delta_2 \exp\left(-\frac{m\Delta_2^2}{4}\right) \\ &\leq \Delta_2 + \frac{4}{\Delta_2}\left(\log\left(\frac{T\Delta_2^2}{4}\right) + 1\right) . \end{aligned}$$
If $m_0 < 1$, then $\Delta_2 < \frac{2}{\sqrt{T}}$. Choosing a random arm yields total expected regret at most
$$\mathcal{R}_T = \frac{T}{2}\Delta_2 \leq \sqrt{T} .$$
If Δ_2 is very small, then we might also just favor the bound
$$\mathcal{R}_T \leq \tfrac{1}{2}\Delta_2 T .$$
Each of these bounds applies in different regimes and tells us different properties of this algorithm. The first bound shows that with appropriate choice of m, explore-then-commit incurs regret asymptotically bounded by $\log(T)$. This is effectively the smallest asymptotic growth achievable and is the gold standard for regret algorithms. However, this logarithmic regret bound depends on the gap Δ_2. For small Δ_2, the second bound shows that the regret is never worse than $\sqrt{T}$ for any value of the gap. $\sqrt{T}$ is one of the more common values for regret, and though it is technically asymptotically worse than logarithmic, algorithms with $\sqrt{T}$ regret tend to be more stable and robust than their logarithmic counterparts. Finally, we note that a very naive algorithm will incur regret that grows linearly with the horizon T. Though linear regret is not typically an ideal situation, there are many applications where it's acceptable. If Δ_2 is tiny to the point where it is hard to observe the difference between μ_1 and μ_2, then linear regret might be completely satisfactory for an application.

2. **Gap-independent regret.** We can also get a gap-independent $\sqrt{T}$ regret for explore then commit. This just requires a bit of calculus:
$$\begin{aligned} \frac{4}{\Delta_2}\left(\log\left(\frac{T\Delta^2}{4}\right) + 1\right) &= 2\sqrt{T}\left(\frac{2}{\Delta_2\sqrt{T}}\left(\log\left(\frac{T\Delta^2}{4}\right) + 1\right)\right) \\ &= 2\sqrt{T} \sup_{x \geq 0} \frac{2\log(x) + 1}{x} \leq 4e^{-1/2}\sqrt{T} \leq 2.5\sqrt{T} . \end{aligned}$$

Hence,

$$\mathcal{R}_T \leq \Delta_2 + 2.5\sqrt{T}$$

no matter what the size of Δ_2 the gap is. Oftentimes this unconditional bound is smaller than the logarithmic bound we derived above.

3. **Gap-independent policy.** The stopping rule we described thus far requires knowing the value of Δ_2. However, if we set $m = T^{2/3}$ then we can achieve sublinear regret no matter what the value of Δ_2 is. To see this again just requires some calculus:

$$\begin{aligned}\mathcal{R}_T &\leq T^{2/3}\Delta_2 + T\Delta_2 \exp\left(-\frac{T^{2/3}\Delta_2^2}{4}\right) \\ &= T^{2/3}\left(\Delta_2 + T^{1/3}\Delta_2 \exp\left(-\frac{T^{2/3}\Delta_2^2}{4}\right)\right) \\ &\leq T^{2/3}\left(\Delta_2 + 2\sup_{x\geq 0} xe^{-x^2}\right) \leq 2T^{2/3}.\end{aligned}$$

$O(T^{2/3})$ regret is technically "worse" than an asymptotic regret of $O(T^{1/2})$, but oftentimes such algorithms perform well in practice. This is because there is a difference between worst-case and average-case behavior, and hence these worst-case bounds on regret themselves do not tell the whole story. A practitioner has to weigh the circumstances of their application to decide what sorts of worst-case scenarios are acceptable.

Interleaving exploration and exploitation

Explore-then-commit is remarkably simple, and illustrates most of the phenomena associated with regret minimization. There are essentially two main shortcomings in the case of the multi-armed bandit:

1. For a variety of practical concerns, it would be preferable to interleave exploration with exploitation.

2. If you don't know the gap, you only get a $T^{2/3}$ rate.

A way to fix this is called *successive elimination*. As in explore-then-commit, we try all actions m times. Then, we drop the actions that are clearly performing poorly. We then try the remaining actions $4m$ times, and drop the poorly performing actions. We run repeated cycles of this pruning procedure, yielding a collection of better actions on average, aiming at convergence to the best return.

Successive Elimination Algorithm:

- Given number of rounds B and an increasing sequence of positive integers $\{m_\ell\}$.
- Initialize the active set of options $\mathcal{A} = \{1, \ldots, K\}$.
- For $\ell = 1, \ldots, B$:

  1. Try every action in $\mathcal{A}$ for m_ℓ times.
  2. Compute the empirical means $\widehat{\mu}_k$ from this iteration only.
  3. Remove from $\mathcal{A}$ any action j with $\mu_j + 2^{-\ell} < \max_{k \in \mathcal{A}} \mu_k$.

The following theorem bounds the regret of successive elimination, and was proven by Auer and Ortner.[224]

Theorem 11. *With $B = \lfloor \frac{1}{2} \log_2 \frac{T}{e} \rfloor$ and $m_\ell = \lceil 2^{2\ell+1} \log \frac{T}{4^\ell} \rceil$, the successive elimination algorithm accrues expected regret*

$$\mathcal{R}_T \leq \sum_{i\colon \Delta_i > \lambda} \left(\Delta_i + \frac{32 \log(T\Delta_i^2) + 96}{\Delta_i} \right) + \max_{i\colon \Delta_i \leq \lambda} \Delta_i T$$

for any $\lambda > \sqrt{e/T}$.

Another popular strategy is known as *optimism in the face of uncertainty*. This strategy is also often called "bet on the best." At iteration t, take all of the observations seen so far and form a set of upper confidence bounds B_i such that

$$\mathbb{P}[\forall i\colon \mu_i \leq B_i(t)] \leq 1 - \delta\,.$$

This leads to the Upper Confidence Bound (UCB) algorithm.

UCB Algorithm

- For $t = 1, \ldots, T$:

  1. Choose action $k = \arg\max_i B_i(t-1)$.
  2. Play action k and observe reward r_t.
  3. Update the confidence bounds.

For the simple case of the multi-armed bandit, we can use the bound that would directly come from Hoeffding's inequality:

$$B_i(t) = \widehat{\mu}_i(t) + \sqrt{\frac{2 \log(1/\delta)}{T_i(t)}}$$

where, we remind the reader, $T_i(t)$ denotes the number of times we have tried action i up to round t. Though $T_i(t)$ is a random variable, one can still prove that this choice yields an algorithm with nearly optimal regret.

More generally, optimistic algorithms work by maintaining an uncertainty set about the dynamics model underlying the SDM problem. The idea is to maintain a set S where we have confidence our true model lies. The algorithm then proceeds by choosing the model in S that gives the highest expected reward. The idea here is that either we get the right model, in which case we get a large reward, or we learn quickly that we have a suboptimal model and we remove it from our set S.

Contextual bandits

Contextual bandits provide a transition from multi-armed bandits to full-fledged reinforcement learning, introducing *state* or *context* into the decision problem. Our goal in contextual bandits is to iteratively update a policy to maximize the total reward:

$$\text{maximize}_{u_t} \ \underset{W_t}{\mathbb{E}} \left[\sum_{t=1}^{T} R(X_t, u_t, W_t) \, . \right]$$

Here, we choose actions u_t according to some policy that is a function of the observations of the random variables X_t, which are called *contexts* or *states*. We make no assumptions about how contexts evolve over time. We assume that the reward function is unknown and, at every time step, the received reward is given by

$$R(X_t, u_t, W_t) = R(X_t, u_t) + W_t$$

where W_t is a random variable with zero mean and independent from all other variables in the problem.

Contextual bandits are a convenient way to abstractly model engagement problems on the internet. In this example, contexts correspond to information about a person. Every interaction a person has with the website can be scored in terms of some sort of reward function that encodes outcomes such as whether the person clicked on an ad, liked an article, or purchased an item. Whatever the reward function is, the goal will be to maximize the total reward accumulated over all time. The X_t will be features describing the person's interaction history, and the action will be related to the content served.

As was the case with the multi-armed bandit, the key idea in solving contextual bandits is to reduce the problem to a prediction problem. In fact we can upper bound our regret by our errors in prediction. The regret accrued by a policy π is

$$\mathbb{E} \left\{ \sum_{t=1}^{T} \max_{u} R(X_t, u) - R(X_t, \pi(X_t)) \right\} .$$

This is because if we know the reward function, then the optimal strategy is to choose the action that maximizes R. This is equivalent to the dynamic programming solution when the dynamics are trivial.

Let's reduce this problem to one of prediction. Suppose that at time t we have built an approximation of the reward function R that we denote $\widehat{R}_t(x,u)$. Let's suppose that our algorithm uses the policy

$$\pi(X_t) = \arg\max_u \widehat{R}_t(X_t, u)\,.$$

That is, we take our current estimate as if it were the true reward function, and pick the action that maximizes reward given the context X_t.

To bound the regret for such an algorithm, note that we have for any action u

$$\begin{aligned} 0 &\le \widehat{R}_t(X_t, \pi(X_t)) - \widehat{R}_t(X_t, u) \\ &\le R(X_t, \pi(X_t)) - R(X_t, u) \\ &+ \left[\widehat{R}_t(X_t, \pi(X_t)) - R(X_t, \pi(X_t))\right] + \left[\widehat{R}_t(X_t, u) - R(X_t, u)\right]\,. \end{aligned}$$

Hence,

$$\sum_{t=1}^{T} \max_u R(X_t, u) - R(X_t, \pi(X_t)) \le 2 \sum_{t=1}^{T} \max_u |\widehat{R}_t(X_t, u) - R(X_t, u)|\,.$$

This final inequality shows that if the prediction error goes to zero, the associated algorithm accrues sublinear regret.

While there are a variety of algorithms for contextual bandits, we focus our attention on two simple solutions that leverage the above reduction to prediction. These algorithms work well in practice and are by far the most commonly implemented. Indeed, they are so common that most applications don't even call these implementations of contextual bandit problems, as they take the bandit nature completely for granted.

Our regret bound naturally suggests the following explore-then-commit procedure.

Explore-then-commit for contextual bandits

- For $t = 1, 2, \ldots, m$:
  1. Receive new context x_t.
  2. Choose a random action u_t.
  3. Receive reward r_t.
- Find a function $\widehat{R}_m$ to minimize prediction error: $\widehat{R}_m := \arg\min_f \sum_{s=1}^{m} loss(f(x_s, u_s), r_s)\,.$
- Define the policy $\pi(x) = \arg\max_u \widehat{R}_m(x_t, u)$.
- For $t = m+1, m+2, \ldots$:
  1. Receive new context x_t.
  2. Choose the action given by $\pi(x_t)$.
  3. Receive reward r_t.

Second, an even more popular method is the following greedy algorithm. The greedy algorithm avoids the initial random exploration stage and instead picks whatever is optimal for the data seen so far.

Greedy algorithm for contextual bandits

- For $t = 1, 2, \dots$:

  1. Find a function $\widehat{R}_t$ to minimize prediction error:

  $$\widehat{R}_t := \arg\min_f \sum_{s=1}^{t-1} loss(f(x_s, u_s), r_s) .$$

  2. Receive new context x_t.
  3. Choose the action given by the policy

  $$\pi_t(x_t) := \arg\max_u \widehat{R}_t(x_t, u) .$$

  4. Receive reward r_t.

In worst-case settings, the greedy algorithm may accrue linear regret. However, worst-case contexts appear to be rare. In the linear contextual bandits problem, where rewards are an unknown linear function of the context, even slight random permutations of a worst-case instance lead to sublinear regret.[225]

The success of the greedy algorithm shows that it is not always desirable to be exploring random actions to see what happens. This is especially true for industrial applications where random exploration is often costly and the value of adding exploration seems limited.[226,227] This context is useful to keep in mind as we move to the more complex problem of reinforcement learning and approximate dynamic programming.

Unknown models and approximate dynamic programming

We now bring dynamics back into the picture and attempt to formalize how to solve general SDM problems when we don't know the dynamics model or even the reward function. We turn to exploring the three main approaches in this space: certainty equivalence fits a model from some collected data and then uses this model as if it were true in the SDM problem. Approximate Dynamic Programming uses Bellman's principle of optimality and stochastic approximation to learn Q-functions from data. Direct Policy Search directly searches for policies by using data from previous episodes in order to improve the reward. Each of these has its advantages and disadvantages as we now explore in depth.

Certainty equivalence for sequential decision making

One of the simplest and perhaps most obvious strategies to solve an SDM problem when the dynamics are unknown is to estimate the dynamics from some data and then to use this estimated model as if it were the true model in the SDM problem.

Estimating a model from data is commonly called "system identification" in the dynamical systems and control literature. System identification differs from conventional estimation because one needs to carefully choose the right inputs to excite various degrees of freedom and because dynamical outputs are correlated over time with the parameters we hope to estimate, the inputs we feed to the system, and the stochastic disturbances. Once data is collected, however, conventional prediction tools can be used to find the system that best agrees with the data and can be applied to analyze the number of samples required to yield accurate models.

Let's suppose we want to build a predictor of the state x_{t+1} from the trajectory history of past observed states and actions. A simple, classic strategy is simply to inject a random probing sequence u_t for control and then measure how the state responds. Up to stochastic noise, we should have that

$$x_{t+1} \approx \varphi(x_t, u_t)\,,$$

where φ is some model aiming to approximate the true dynamics. φ might arise from a first-principles physical model or might be a non-parametric approximation by a neural network. The state-transition function can then be fit using supervised learning. For instance, a model can be fit by solving the least-squares problem

$$\text{minimize}_{\varphi} \quad \textstyle\sum_{t=0}^{N-1} ||x_{t+1} - \varphi(x_t, u_t)||^2\,.$$

Let $\widehat{\varphi}$ denote the function fit to the collected data to model the dynamics. Let ω_t denote a random variable that we will use as a model for the noise process. With such a point estimate for the model, we might solve the optimal control problem

$$\begin{array}{ll} \text{maximize} & \mathbb{E}_{\omega_t}[\sum_{t=0}^{N} R(x_t, u_t)] \\ \text{subject to} & x_{t+1} = \widehat{\varphi}(x_t, u_t) + \omega_t,\ u_t = \pi_t(\tau_t)\,. \end{array}$$

In this case, we are solving the wrong problem to get our control policies π_t. Not only is the model incorrect, but this formulation requires some plausible model of the noise process. But we emphasize that this is standard engineering practice. Though more sophisticated techniques can be used to account for the errors in modeling, feedback often can compensate for these modeling errors.

Approximate dynamic programming

Approximate dynamic programming approaches the RL problem by directly approximating the optimal control cost and then solving this with techniques from dynamic programming. Approximate Dynamic Programming methods typically try to infer Q-functions directly from data. The standard assumption in most practical implementations of Q-learning is that the Q-functions are static, as would be the case in the infinite horizon, discounted optimal control problem.

Probably the best known approximate dynamic programming method is *Q-learning*.[228] Q-learning simply attempts to solve value iteration using *stochastic approximation*. If we draw a sample trajectory using the policy given by the optimal policy, then we should have (approximately and in expectation)

$$\mathcal{Q}_\gamma(x_t, u_t) \approx R(x_t, u_t) + \gamma \max_{u'} \mathcal{Q}_\gamma(x_{t+1}, u')\,.$$

Thus, beginning with some initial guess $\mathcal{Q}_\gamma^{(\text{old})}$ for the Q-function, we can update

$$\mathcal{Q}_\gamma^{(\text{new})}(x_t, u_t) = (1-\eta)\mathcal{Q}_\gamma^{(\text{old})}(x_t, u_t) + \eta \left(R(x_t, u_t) + \gamma \max_{u'} \mathcal{Q}_\gamma^{(\text{old})}(x_{t+1}, u') \right)$$

where η is a *step-size* or *learning rate*.

The update here only requires data generated by the policy Q_γ^{old} and does not need to know the explicit form of the dynamics. Moreover, we don't even need to know the reward function if this is provided online when we generate trajectories. Hence, Q-learning is often called "model free." We strongly dislike this terminology and do not wish to dwell on it. Unfortunately, distinguishing between what is "model-free" and what is "model-based" tends to just lead to confusion. All reinforcement learning is inherently based on models, as it implicitly assumes data is generated by some Markov Decision Process. In order to run Q-learning we need to know the form of the Q-function itself, and except for the tabular case, how to represent this function requires *some* knowledge of the underlying dynamics. Moreover, assuming that value iteration is the proper solution of the problem is a modeling assumption: we are assuming a discount factor and time invariant dynamics. But the reader should be advised that when they read "model-free," this almost always means "no model of the state transition function was used when running this algorithm."

For continuous control problems, methods like Q-learning appear to make an inefficient use of samples. Suppose the internal state of the system is of dimension d. When modeling the state-transition function, each sample provides d pieces of information about the dynamics. By contrast, Q-learning uses only one piece of information per time step. Such inefficiency is often seen in practice. Also troubling is the fact that we had to introduce the discount factor in order to get a simple form of the Bellman equation. One can avoid discount factors, but this requires considerably more sophisticated analysis. Large discount factors do in practice lead to brittle methods, and the discount factor becomes a hyperparameter that must be tuned to stabilize performance.

We close this section by noting that for many problems with high-dimensional states or other structure, we might be interested in not representing Q-functions as a lookup table. Instead, we might approximate the Q-functions with a parametric family: $\mathcal{Q}(x,u;\vartheta)$. Though we'd like to update the parameter ϑ using something like gradient descent, it's not immediately obvious how to do so. The simplest attempt, following the guide of stochastic approximation, is to run the iterations:

$$\begin{aligned}\delta_t &= R(x_t,u_t) + \gamma \mathcal{Q}(x_{t+1},u_{t+1};\vartheta_t) - \mathcal{Q}(x_t,u_t;\vartheta_t)\\ \vartheta_{t+1} &= \vartheta_t + \eta\delta_t \nabla \mathcal{Q}(x_t,u_t,\vartheta_t)\,.\end{aligned}$$

This algorithm is called *Q-learning with function approximation*. A typically more stable version uses momentum to average out noise in Q-learning. With δ_t as above, we add the modification

$$\begin{aligned}e_t &= \lambda e_{t-1} + \nabla \mathcal{Q}(x_t,u_t,\vartheta_t)\\ \vartheta_{t+1} &= \vartheta_t + \eta\delta_t e_t\end{aligned}$$

for $\lambda \in [0,1]$. This method is known as SARSA(λ).[229]

Direct policy search

The most ambitious form of control without models attempts to directly learn a policy function from episodic experiences without ever building a model or appealing to the Bellman equation. From the oracle perspective, these policy driven methods turn the problem of RL into derivative-free optimization.

In turn, let's first begin with a review of a general paradigm for leveraging random sampling to solve optimization problems. Consider the general unconstrained optimization problem

$$\text{maximize}_{z\in\mathbb{R}^d} \quad R(z)\,.$$

Any optimization problem like this is equivalent to an optimization over probability densities on z:

$$\text{maximize}_{p(z)} \quad \mathbb{E}_p[R(z)]\,.$$

If $z_\star$ is the optimal solution, then we'll get the same value if we put a δ-function around $z_\star$. Moreover, if p is a density, it is clear that the *expected value of the reward function* can never be larger than the maximal reward achievable by a fixed z. So we can either optimize over z or we can optimize over *densities* over z.

Since optimizing over the space of all probability densities is intractable, we must restrict the class of densities over which we optimize. For example, we can consider a family parameterized by a parameter vector ϑ: $p(z;\vartheta)$ and attempt to optimize

$$\text{maximize}_{\vartheta} \quad \mathbb{E}_{p(z;\vartheta)}[R(z)] .$$

If this family of densities contains all of the Delta functions, then the optimal value will coincide with the nonrandom optimization problem. But if the family does not contain the Delta functions, the resulting optimization problem only provides a lower bound on the optimal value no matter how good of a probability distribution we find.

That said, this reparameterization provides a powerful and general algorithmic framework for optimization. In particular, we can compute the derivative of $J(\vartheta) := \mathbb{E}_{p(z;\vartheta)}[R(z)]$ using the following calculation (called "the log-likelihood trick"):

$$\begin{aligned}
\nabla_{\vartheta} J(\vartheta) &= \int R(z) \nabla_{\vartheta} p(z;\vartheta)\, \mathrm{d}z \\
&= \int R(z) \left(\frac{\nabla_{\vartheta} p(z;\vartheta)}{p(z;\vartheta)} \right) p(z;\vartheta)\, \mathrm{d}z \\
&= \int \left(R(z) \nabla_{\vartheta} \log p(z;\vartheta) \right) p(z;\vartheta)\, \mathrm{d}z \\
&= \mathbb{E}_{p(z;\vartheta)} \left[R(z) \nabla_{\vartheta} \log p(z;\vartheta) \right] .
\end{aligned}$$

This derivation reveals that the gradient of J with respect to ϑ is the expected value of the function

$$G(z,\vartheta) = R(z) \nabla_{\vartheta} \log p(z;\vartheta) .$$

Hence, if we sample z from the distribution defined by $p(z;\vartheta)$, we can compute $G(z,\vartheta)$ and will have an unbiased estimate of the gradient of J. We can follow this direction and will be running stochastic gradient descent on J, defining the following algorithm:

REINFORCE algorithm:

- *Input Hyperparameters:* step-sizes $\alpha_j > 0$.
- *Initialize:* ϑ_0 and $k = 0$.
- Until the heat death of the universe, do:

    1. Sample $z_k \sim p(z;\vartheta_k)$.
    2. Set $\vartheta_{k+1} = \vartheta_k + \alpha_k R(z_k) \nabla_{\vartheta} \log p(z_k;\vartheta_k)$.
    3. $k \leftarrow k + 1$.

The main appeal of the REINFORCE Algorithm is that it is not hard to implement. If you can efficiently sample from $p(z;\vartheta)$ and can easily compute $\nabla \log p$, you can run this algorithm on essentially any problem. But such generality must and does come with a significant cost. The algorithm operates on stochastic gradients of the sampling distribution, but the function we cared about optimizing—R—is only accessed through function evaluations. Direct search methods that use the log-likelihood trick are necessarily derivative-free optimization methods, and, in turn, are necessarily less effective than methods that compute actual gradients, especially when the function evaluations are noisy. Another significant concern is that the choice of distribution can lead to very high variance in the stochastic gradients. Such high variance in turn implies that many samples need to be drawn to find a stationary point.

That said, the ease of implementation should not be readily discounted. Direct search methods are easy to implement, and oftentimes reasonable results can be achieved with considerably less effort than custom solvers tailored to the structure of the optimization problem. There are two primary ways that this sort of stochastic search arises in reinforcement learning: Policy gradient and pure random search.

Policy gradient

Though we have seen that the optimal solutions of Bellman's equations are deterministic, probabilistic policies can add an element of exploration to a control strategy, hopefully enabling an algorithm to simultaneously achieve reasonable awards and learn more about the underlying dynamics and reward functions. Such policies are the starting point for *policy gradient* methods.[230]

Consider a *parametric, randomized policy* such that u_t is sampled from a distribution $p(u|\tau_t;\vartheta)$ that is a function only of the currently observed trajectory and a parameter vector ϑ. A probabilistic policy induces a probability distribution over trajectories:

$$p(\tau;\vartheta) = \prod_{t=0}^{L-1} p(x_{t+1}|x_t,u_t)p(u_t|\tau_t;\vartheta)\,.$$

Moreover, we can overload notation and define the reward of a trajectory to be

$$R(\tau) = \sum_{t=0}^{N} R_t(x_t,u_t)\,.$$

Then our optimization problem for reinforcement learning takes the form of stochastic search. Policy gradient thus proceeds by sampling a trajectory using the probabilistic policy with parameters ϑ_k, and then updating using REINFORCE.

Using the log-likelihood trick and the factored form of the probability distribution $p(\tau;\vartheta)$, we can see that the gradient of J with respect to ϑ *is not a function of the underlying dynamics*. However, at this point this should not be surprising: by shifting to distributions over policies, we push the burden of optimization onto the sampling procedure.

Pure random search

An older and more widely applied method to solve the generic stochastic search problem is to directly perturb the current decision variable z by random noise and then update the model based on the received reward at this perturbed value. That is, we apply the REINFORCE Algorithm with sampling distribution $p(z;\vartheta) = p_0(z-\vartheta)$ for some distribution p_0. Simplest examples for p_0 would be the uniform distribution on a sphere or a normal distribution. Perhaps less surprisingly here, REINFORCE can again be run without any knowledge of the underlying dynamics. The REINFORCE algorithm has a simple interpretation in terms of gradient approximation. Indeed, REINFORCE is equivalent to approximate gradient ascent of R

$$\vartheta_{t+1} = \vartheta_t + \alpha g_\sigma(\vartheta_k)$$

with the gradient approximation

$$g_\sigma(\vartheta) = \frac{R(\vartheta+\sigma\epsilon) - R(\vartheta-\sigma\epsilon)}{2\sigma}\epsilon\,.$$

This update asks to compute a finite-difference approximation to the gradient along the direction ϵ and move along the gradient. One can reduce the variance of such a finite-difference estimate by sampling along multiple random directions and averaging:

$$g_\sigma^{(m)}(\vartheta) = \frac{1}{m}\sum_{i=1}^{m} \frac{R(\vartheta+\sigma\epsilon_i) - R(\vartheta-\sigma\epsilon_i)}{2\sigma}\epsilon_i\,.$$

This is akin to approximating the gradient in the random subspace spanned by the ϵ_i

This particular algorithm and its generalizations go by many different names. Probably the earliest proposal for this method is by Rastrigin.[231] Somewhat surprisingly, Rastrigin initially developed this method to solve reinforcement learning problems. His main motivating example was an inverted pendulum. A rigorous analysis using contemporary techniques was provided by Nesterov and Spokoiny.[232] Random search was also discovered by the evolutionary algorithms community and is called (μ,λ)-Evolution Strategies.[233,234] Random search has also been studied in the context of stochastic approximation[235] and bandits.[236,237] Algorithms that get invented by four different communities probably have something good going for them.

Deep reinforcement learning

We have thus far spent no time discussing *deep* reinforcement learning. That is because there is nothing conceptually different other than using neural networks for function approximation. That is, if one wants to take any of the described methods and make them deep, they simply need to add a neural net. In model-based RL, φ is parameterized as a neural net. In ADP the Q-functions or Value Functions are assumed to be well approximated by neural nets. In policy search, the policies are set to be neural nets. The algorithmic concepts themselves don't change. However, convergence analysis certainly will change, and algorithms like Q-learning might not even converge. The classic text *Neurodynamic Programming* by Bertsekas and Tsitisklis discusses the adaptations needed to admit function approximation.[238]

Certainty equivalence is often optimal

In this section, we give a survey of the power of certainty equivalence in sequential decision making problems. We focus on the simple cases of tabular MDP and LQR as they are illustrative of more general problems while still being manageable enough to analyze with relatively simple mathematics. However, these analyses are less than a decade old. Though the principle of certainty equivalence dates back over 60 years, our formal understanding of certainty equivalence and its robustness is just now solidifying.

Certainty equivalence for LQR

Consider the linear quadratic regulator problem

$$\begin{array}{ll} \text{minimize} & \lim_{T\to\infty} \mathbb{E}_{W_t}\left[\frac{1}{2T}\sum_{t=0}^{T} X_t^T\Phi X_t + U_t^T\Psi U_t\right], \\ \text{subject to} & X_{t+1} = AX_t + BU_t + W_t,\ U_t = \pi_t(X_t) \\ & (x_0 \text{ given}). \end{array}$$

We have shown that the solution to this problem is static state feedback $U_t = -K_\star X_t$ where

$$K_\star = (\Psi + B^T MB)^{-1} B^T MA$$

and M is the unique stabilizing solution to the Discrete Algebraic Riccati Equation

$$M = \Phi + A^T MA - (A^T MB)(\Psi + B^T MB)^{-1}(B^T MA)\,.$$

Suppose that instead of knowing (A, B, Φ, Ψ) exactly, we only have estimates $(\widehat{A}, \widehat{B}, \widehat{\Phi}, \widehat{\Psi})$. Certainty equivalence would then yield a control policy $U_t = -\widehat{K}X_t$ where $\widehat{K}$ can be found using $(\widehat{A}, \widehat{B}, \widehat{\Phi}, \widehat{\Psi})$ in place of (A, B, Φ, Ψ) in the formulae above. What is the cost of this model?

The following discussion follows arguments due to Mania et al.[239] Let $J(K)$ denote the cost of using the policy K. Note that this cost may be infinite, but it will also be differentiable in K. If we unroll the dynamics and compute expected values, one can see that the cost is the limit of polynomials in K, and hence is differentiable.

Suppose that

$$\epsilon := \max\left\{\|\widehat{A} - A\|, \|\widehat{B} - B\|, \|\widehat{\Phi} - \Phi\|, \|\widehat{\Psi} - \Psi\|\right\}$$

If we Taylor expand the cost we find that for some $t \in [0,1]$,

$$J(\widehat{K}) - J_\star = \langle \nabla J(K_\star), \widehat{K} - K_\star \rangle + \frac{1}{2}(\widehat{K} - K_\star)^T \nabla^2 J(\tilde{K})(\widehat{K} - K_\star)$$

where $\tilde{K} = (1-t)K_\star + t\widehat{K}$. The first term is equal to zero because $K_\star$ is optimal. Since the map from (A, B, Φ, Ψ) to $K_\star$ is differentiable, there must be constants L and ϵ_0 such that $\|\widehat{K} - K_\star\| \leq L\epsilon$ whenever $\epsilon \leq \epsilon_0$. This means that as long as the estimates for (A, B, Φ, Ψ) are close enough to the true values, we must have

$$J(\widehat{K}) - J_\star = O(\epsilon^2)\,.$$

Just how good should the estimates for these quantities be? Let's focus on the dynamics (A, B) as the cost matrices Φ and Ψ are typically design parameters, not unknown properties of the system. Suppose A is $d \times d$ and B is $d \times p$. Basic parameter counting suggests that if we observe T sequential states from the dynamical system, we observe a total of dT numbers, one for each dimension of the state per time step. Hence, a naive statistical guess would suggest that

$$\max\left\{\|\widehat{A} - A\|, \|\widehat{B} - B\|\right\} \leq O\left(\sqrt{\frac{d+p}{T}}\right).$$

Combining this with our Taylor series argument implies that

$$J(\widehat{K}) - J_\star = O\left(\frac{d+p}{T}\right).$$

As we already described, this also suggests that certainty equivalent control accrues a regret of

$$\mathcal{R}_T = O(\sqrt{T})\,.$$

This argument can be made completely rigorous.[239] The regret accrued also turns out to be the optimal.[240] Moreover, the Taylor series argument here works for any model where the cost function is twice differentiable. Thus, we'd expect to see similar behavior in more general SDM problems with continuous state spaces.

Certainty equivalence for tabular MDPs

For discounted, tabular MDPs, certainty equivalence also yields an optimal sample complexity. This result is elementary enough to be proven in a few pages. We first state an approximation theorem that shows that if you build a policy with the wrong model, the value of that policy can be bounded in terms of the inaccuracy of your model. Then, using Hoeffding's inequality, we can construct a sample complexity bound that is nearly optimal. The actual optimal rate follows using our main approximation theorem coupled with slightly more refined concentration inequalities. We refer readers interested in this more refined analysis to the excellent reinforcement learning text by Agarwal et al.[241]

Let $V_\star(x)$ denote the optimal expected reward attainable by some policy on the discounted problem

$$\begin{array}{ll} \text{maximize} & (1-\gamma)\,\mathbb{E}_{W_t}[\sum_{t=0}^{\infty} \gamma^t R(X_t, U_t, W_t)] \\ \text{subject to} & X_{t+1} = f(X_t, U_t, W_t),\ U_t = \pi_t(X_t) \\ & (x_0 = x). \end{array}$$

Note that $V_\star$ is a function of the initial state. This mapping from initial state to expected rewards is called the *value function* of the SDM problem. Let $V^\pi(x)$ denote the expected reward attained when using some fixed, static policy π. Our aim is to evaluate the reward of particular policies that arise from certainty equivalence.

To proceed, first let $\widehat{Q}$ be any function mapping state-action pairs to a real value. We can always define a policy $\pi_{\widehat{Q}}(x) = \arg\max_u \widehat{Q}(x, u)$. The following theorem quantifies the value of $\pi_{\widehat{Q}}$ when $\widehat{Q}$ is derived by solving the Bellman equation with an approximate model of the MDP dynamics. This theorem has been derived in numerous places in the RL literature, and yet it does not appear to be particularly well known. As pointed out by Ávila Pires and Szepesvari,[242] it appears as a corollary to Theorem 3.1 in Whitt[243] (1978), as Corollary 2 in Singh and Yee[244] (1994), and as a corollary of Proposition 3.1 in Bertsekas[245] (2012). We emphasize it here as it demonstrates immediately why certainty equivalence is such a powerful tool in sequential decision making problems.

Theorem 12. ***Model-error for MDPs.*** *Consider a γ-discounted MDP with dynamics governed by a model p and a reward function r. Let $\widehat{Q}$ denote the Q-function for the MDP with the same rewards but dynamics $\widehat{\mathbb{P}}$. Then we have*

$$V_\star(x) - V^{\pi_{\widehat{Q}}}(x) \leq \frac{2\gamma}{(1-\gamma)^2} \sup_{x,u} \left| \mathop{\mathbb{E}}_{\widehat{\mathbb{P}}(\cdot|x,u)}[V_\star] - \mathop{\mathbb{E}}_{\mathbb{P}(\cdot|x,u)}[V_\star] \right| .$$

This theorem states that the values associated with the policy that we derive using the wrong dynamics will be close to optimal if $\mathbb{E}_{\widehat{\mathbb{P}}(\cdot|x,u)}[V_\star]$ and $\mathbb{E}_{\mathbb{P}(\cdot|x,u)}[V_\star]$ are close for all state-action pairs (x, u). This is a remarkable result as it shows that we can control our regret by our prediction errors.

But the only prediction that matters is our predictions of the optimal value vectors $V_\star$. Note further that this theorem makes no assumptions about the size of the state spaces: it holds for discrete, tabular MDPs, and more general discounted MDPs. A discussion of more general problems is covered by Bertsekas.[246]

Focusing on the case of finite-state tabular MDPs, suppose the rewards are in the range $[0,1]$ and there are S states and A actions. Then the values are in the range $V_\star(x) \in [0, (1-\gamma)^{-1}]$. Immediately from this result, we can derive a sample-complexity bound. Let's suppose that for each pair (x,u), we collect n samples to estimate the conditional probabilities $\mathbb{P}[X' = x'|x,u]$, and define $\widehat{\mathbb{P}}(X' = x'|x,u)$ to be the number of times we observe x' divided by n. Then by Hoeffding's inequality

$$\mathbb{P}\left[\left|\underset{\widehat{\mathbb{P}}[\cdot|x,u]}{\mathbb{E}}[V_\star] - \underset{\mathbb{P}[\cdot|x,u]}{\mathbb{E}}[V_\star]\right| \geq \epsilon\right] \leq 2\exp\left(-2n\epsilon^2(1-\gamma)^2\right)$$

and therefore, by the union bound

$$\sup_{x,u}\left|\underset{\widehat{\mathbb{P}}[\cdot|x,u]}{\mathbb{E}}[V_\star] - \underset{\mathbb{P}[\cdot|x,u]}{\mathbb{E}}[V_\star]\right| \leq \sqrt{\frac{\log\left(\frac{2SA}{\delta}\right)}{n(1-\gamma)^2}}$$

with probability $1-\delta$. If we let $N = SAn$ denote the total number of samples collected, we see that

$$V_\star(x) - V^{\pi_{\widehat{Q}}}(x) \leq \frac{2\gamma}{(1-\gamma)^3}\sqrt{\frac{SA\log\left(\frac{2SA}{\delta}\right)}{N}}\,.$$

Our naive bound here is nearly optimal: the dependence on γ can be reduced from $(1-\gamma)^{-3}$ to $(1-\gamma)^{-3/2}$ using more refined deviation inequalities, but the dependence on S, A, and N is optimal.[241,247] That is, certainty equivalence achieves an optimal sample complexity for the discounted tabular MDP problem.

Proof of the model-error theorem

The proof here combines arguments by Bertsekas[245] and Agarwal.[241] Let us first introduce notation that makes the proof a bit more elegant. Let Q be any function mapping state-action pairs to real numbers. Given a policy π and a state-transition model $\mathbb{P}$, denote $\mathcal{T}_{\mathbb{P}}$ to be the map from functions to functions where

$$[\mathcal{T}_{\mathbb{P}}Q](x,u) = r(x,u) + \gamma\sum_{x'}\max_{u'} Q(x',u')\,\mathbb{P}[X' = x'|x,u]\,.$$

With this notation, the Bellman equation for the discounted MDP simply becomes

$$Q_\star = \mathcal{T}_{\mathbb{P}} Q_\star \,.$$

If we were to use $\widehat{\mathbb{P}}$ instead of $\mathbb{P}$, this would yield an alternative Q-function, $\widehat{Q}$, that satisfies the Bellman equation $\widehat{Q} = \mathcal{T}_{\widehat{\mathbb{P}}}\widehat{Q}$.

The operator $\mathcal{T}_{\mathbb{P}}$ is a *contraction mapping* in the ℓ_∞ norm. To see this, note that for any functions Q_1 and Q_2,

$$\begin{aligned}
|[\mathcal{T}_{\mathbb{P}} Q_1 - \mathcal{T}_{\mathbb{P}} Q_2](x,u)| &= \left|\gamma \sum_{x'} (\max_{u_1} Q_1(x',u_1) - \max_{u_2} Q_2(x',u_2))\, \mathbb{P}[X' = x'|x,u]\right| \\
&\le \gamma \sum_{x'} \mathbb{P}[X' = x'|x,u] \left|\max_{u_1} Q_1(x',u_1) - \max_{x_2} Q_2(x',u_2)\right| \\
&\le \gamma \|Q_1 - Q_2\|_\infty \,.
\end{aligned}$$

Since $\mathcal{T}_{\mathbb{P}}$ is a contraction, the solution of the discounted Bellman equations are unique and $Q_\star = \lim_{k\to\infty} \mathcal{T}_{\mathbb{P}}{}^k Q$ for any function Q. Similarly, $\widehat{Q} = \lim_{k\to\infty} \mathcal{T}_{\widehat{\mathbb{P}}}{}^k Q$.

Now we can bound

$$\left\|\mathcal{T}_{\widehat{\mathbb{P}}}{}^k Q_\star - Q_\star\right\|_\infty \le \sum_{i=1}^k \|\mathcal{T}_{\widehat{\mathbb{P}}}{}^i Q_\star - \mathcal{T}_{\widehat{\mathbb{P}}}{}^{i-1} Q_\star\|_\infty \le \sum_{i=1}^k \gamma^k \|\mathcal{T}_{\widehat{\mathbb{P}}} Q_\star - Q_\star\|_\infty \,.$$

Taking limits of both sides as $k \to \infty$, we find

$$\left\|\widehat{Q} - Q_\star\right\|_\infty \le \frac{1}{1-\gamma} \|\mathcal{T}_{\widehat{\mathbb{P}}} Q_\star - Q_\star\|_\infty \,.$$

But since $Q_\star = \mathcal{T}_{\mathbb{P}} Q_\star$, and

$$\begin{aligned}
&[\mathcal{T}_{\widehat{\mathbb{P}}} Q_\star - \mathcal{T}_{\mathbb{P}} Q_\star](x,u) \\
=&\gamma \sum_{x'} \max_{u'} Q_\star(x',u') \left(\widehat{\mathbb{P}}[X' = x'|x,u] - \mathbb{P}[X' = x'|x,u]\right) ,
\end{aligned}$$

we have the bound

$$\left\|\widehat{Q} - Q_\star\right\|_\infty \le \frac{\gamma}{1-\gamma} \sup_{x,u} \left| \mathop{\mathbb{E}}_{\widehat{\mathbb{P}}[\cdot|x,u]}[V_\star] - \mathop{\mathbb{E}}_{\mathbb{P}[\cdot|x,u]}[V_\star] \right| \,.$$

To complete the proof, it suffices to use $\left\|\widehat{Q} - Q_\star\right\|_\infty$ to upper bound the difference between the values of the two policies. Let $\pi_\star(x) = \arg\max_u Q_\star(x,u)$ and $\widehat{\pi}(x) = \arg\max_u \widehat{Q}_\star(x,u)$ denote the optimal policies for the models $\mathbb{P}$ and $\widehat{\mathbb{P}}$ respectively. For any policy π, we have

$$V^\pi(x) = r(x,\pi(x)) + \gamma \sum_{x'} \mathbb{P}[X' = x'|x,\pi(x)] V^\pi(x') \,,$$

and hence we can bound the optimality gap as

$$\begin{aligned}
V_\star(x) - V^{\widehat{\pi}}(x) &= Q_\star(x, \pi_\star(x)) - V^{\widehat{\pi}}(x) \\
&= Q_\star(x, \pi_\star(x)) - Q_\star(x, \widehat{\pi}(x)) + Q_\star(x, \widehat{\pi}(x)) - V^{\widehat{\pi}}(x) \\
&= Q_\star(x, \pi_\star(x)) - Q_\star(x, \widehat{\pi}(x)) \\
&\qquad + \gamma \sum_{x'} \mathbb{P}[X' = x' | x, \widehat{\pi}(x)] \left(V_\star(x') - V^{\widehat{\pi}}(x') \right) \\
&\leq Q_\star(x, \pi_\star(x)) - \widehat{Q}(x, \pi_\star(x)) + \widehat{Q}(x, \widehat{\pi}(x)) - Q_\star(x, \widehat{\pi}(x)) \\
&\qquad + \gamma \sum_{x'} \mathbb{P}[X' = x' | x, \widehat{\pi}(x)] \left(V_\star(x') - V^{\widehat{\pi}}(x') \right) \\
&\leq 2\|Q_\star - \widehat{Q}\|_\infty + \gamma \|V_\star - V^{\widehat{\pi}}\|_\infty .
\end{aligned}$$

Here, the first inequality holds because $\widehat{Q}(x, \pi_\star(x)) \leq \max_u \widehat{Q}(x, u) = \widehat{Q}(x, \widehat{\pi}(x))$. Rearranging terms shows that

$$V_\star(x) - V^{\widehat{\pi}}(x) \leq \frac{2}{1-\gamma} \|Q_\star - \widehat{Q}\|_\infty ,$$

which, when combined with our previous bound on $\|Q_\star - \widehat{Q}\|_\infty$, completes the proof.

Sample complexity of other RL algorithms

The sample complexity of reinforcement learning remains an active field, with many papers honing in on algorithms with optimal complexity. Researchers have now shown a variety of methods achieve the optimal complexity of certainty equivalence, including those based on ideas from approximate dynamic programming and Q-learning. For LQR on the other hand, no other methods are currently competitive.

While sample complexity is important, there are no significant gains to be made over simple baselines that echo decades of engineering practice. And, unfortunately, though sample complexity is a well-posed problem that excites many researchers, it does not address many of the impediments preventing reinforcement learning from being deployed in more applications. As we will see in a moment, the optimization framework itself has inherent weaknesses that cannot be fixed by better sample efficiency, and these weaknesses must be addressed head-on when designing an SDM system.

The limits of learning in feedback loops

Though we have shown the power of certainty equivalence, it is also a useful example to guide how reinforcement learning—and optimal sequential decision making more generally—can go wrong. First, we will show how optimal decision making problems themselves can be set up to be very sensitive to model-error. So treating a model as true in these cases can lead to misguided optimism about performance. Second, we will adapt this example to the case where the state is partially observed and demonstrate a more subtle pathology. As we discussed in the last chapter, when state is not perfectly observed, decision making is decidedly more difficult. Here we will show an example where improving your prediction paradoxically increases your sensitivity to model-error.

Fragile instances of the linear quadratic regulator

Consider the following innocuous dynamics:

$$A = \begin{bmatrix} 0 & 1 \\ 0 & 0 \end{bmatrix}, \quad B = \begin{bmatrix} 0 \\ 1 \end{bmatrix}.$$

This system is a simple, two-state shift register. Write the state out with indexed components $x = [x^{(1)}, x^{(2)}]^\top$. New states enter through the control B into the second state. The first state $x^{(1)}$ is simply whatever was in the second register at the previous time step. The open loop dynamics of this system are as stable as you could imagine. Both eigenvalues of A are zero.

Let's say our control objective aims to try to keep the two components of the state equal to each other. We can model this with the quadratic cost matrices

$$\Phi = \begin{bmatrix} 1 & -1 \\ -1 & 1 \end{bmatrix}, \quad \Psi = 0.$$

Here, $\Psi = 0$ for simplicity, as the formulas are particularly nice for this case. But, as we will discuss in a moment, the situation is not improved simply by having R be positive. For the disturbance, assume that W_t is zero-mean, has bounded second moment, $\Sigma_t = \mathbb{E}[W_t W_t^\top]$, and is uncorrelated with X_t and U_t.

The cost is asking to minimize

$$\mathbb{E}\left[\sum_{t=1}^{N} (X_t^{(1)} - X_t^{(2)})^2.\right]$$

When $W_t = 0$, $X_t^{(1)} + X_t^{(2)} = X_{t-1}^{(2)} + U_{t-1}$, so the intuitive best action would be to set $U_t = X_t^{(2)}$. This turns out to be the optimal action, and one can prove this directly using standard dynamic programming computations or a Discrete Algebraic Riccati Equation. With this identification, we can write *closed loop* dynamics by eliminating the control signal:

$$X_{t+1} = \begin{bmatrix} 0 & 1 \\ 0 & 1 \end{bmatrix} X_t + W_t.$$

This closed loop system is *marginally stable*, meaning that while signals don't blow up, some states will persist forever and not converge to 0. The second component of the state simply exhibits a random walk on the real line. We can analytically see that the system is not stable by computing the eigenvalues of the state-transition matrix, which are here 0 and 1. The 1 corresponds the state where the two components are equal, and such a state can persist forever.

If we learned an incorrect model of the dynamics, how would that influence the closed loop behavior? The simplest scenario is that we identified B from some preliminary experiments. If the true $B_\star = \alpha B$, then the closed loop dynamics are

$$X_{t+1} = \begin{bmatrix} 0 & 1 \\ 0 & \alpha \end{bmatrix} X_t + W_t \,.$$

This system is unstable for any $\alpha > 1$. That is, the system is arbitrarily sensitive to misidentification of the dynamics. This lack of robustness has nothing to do with the noise sequence. The structure of the cost is what drives the system to fragility.

If $\Psi > 0$, we would get a slightly different policy. Again, using elementary dynamic programming shows that the optimal control is $u_t = \beta_t(\Psi) x_t^{(2)}$ for some $\beta_t(\Psi) \in (1/2, 1)$. The closed loop system will be a bit more stable, but this comes at the price of reduced performance. You can also check that if you add ϵ times the identity to Φ, we again get a control policy proportional to the second component of the state, $x_t^{(2)}$.

Similar examples are fairly straightforward to construct. The state-transition matrix of the closed loop dynamics will always be of the form $A - BK$, and we can first find a K such that $A - BK$ has an eigenvalue of magnitude 1. Once this is constructed, it suffices to find a vector v such that $(v'B)^{-1}v'A = K$. Then the cost $\Phi = vv'$ yields the desired pathological example.

One such example that we will use in our discussion of partially observed systems is the model:

$$A = \begin{bmatrix} 1 & 1 \\ 0 & 1 \end{bmatrix}, \quad B = \begin{bmatrix} 0 \\ 1 \end{bmatrix},$$

$$\Phi = \begin{bmatrix} 1 & 1/2 \\ 1/2 & 1/4 \end{bmatrix}, \quad \Psi = 0\,.$$

The dynamics here are our "Newton's Law" dynamics studied in our dynamic programming examples. One can check that the closed loop dynamics of this system are

$$X_{t+1} = \begin{bmatrix} 1 & 1 \\ -2 & -2 \end{bmatrix} X_t + W_t \,.$$

The transition matrix here has eigenvalues 0 and -1, and the state $x = [1/2, -1]$ will oscillate in sign and persist forever.

Partially observed example

Recall that the generalization of LQR to the case with imperfect state observation is called "Linear Quadratic Gaussian" control (LQG). This is the simplest, special case of a POMDP. We again assume linear dynamics:

$$X_{t+1} = AX_t + BU_t + W_t\,,$$

where the state is now corrupted by zero-mean Gaussian noise, W_t. Instead of measuring the state X_t directly, we instead measure a signal Y_t of the form

$$Y_t = CX_t + V_t\,.$$

Here, V_t is also zero-mean Gaussian noise. Suppose we'd still like to minimize a quadratic cost function

$$\lim_{T\to\infty} \mathbb{E}\left[\frac{1}{T}\sum_{t=0}^{T} X_t^\top \Phi X_t + U_t^\top \Psi U_t\right]\,.$$

This problem is very similar to our LQR problem except for the fact that we get an indirect measurement of the state and need to apply some sort of *filtering* of the Y_t signal to estimate X_t.

The optimal solution for LQG is strikingly elegant. Since the observation of X_t is through a Gaussian process, the maximum likelihood estimation algorithm has a clean, closed form solution. As we saw in the previous chapter, our best estimate for X_t, denoted $\widehat{x}_t$, given all of the data observed up to time t, is given by a Kalman Filter. The estimate obeys a difference equation

$$\widehat{x}_{t+1} = A\widehat{x}_t + Bu_t + L(y_t - C\widehat{x}_t)\,.$$

The matrix L can be found by solving a discrete algebraic Riccati equation that depends on the variance of v_t and w_t and on the matrices A and C. In particular, it's the DARE with data $(A^\top, C^\top, \Sigma_w, \Sigma_v)$.

The optimal LQG solution takes the estimate of the Kalman Filter, $\widehat{x}_t$, and sets the control signal to be

$$u_t = -K\widehat{x}_t\,.$$

Here, K is gain matrix that would be used to solve the LQR problem with data (A, B, Φ, Ψ). That is, LQG performs optimal filtering to compute the best state estimate, and then computes a feedback policy as if this estimate were a noiseless measurement of the state. That this turns out to be optimal is one of the more amazing results in control theory. It decouples the process of designing an optimal filter from designing an optimal controller, enabling simplicity and modularity in control design. This decoupling where we treat the output of our state estimator as the true state is yet another example of certainty equivalence, and yet another example of where certainty equivalence turns out to be optimal. However, as we will now see, LQG highlights a particular scenario where certainty equivalent control leads to misplaced optimism about robustness.

Before presenting the example, let's first dive into *why* LQG is likely less robust than LQR. Let's assume that the true dynamics are generated as:

$$X_{t+1} = AX_t + B_\star U_t + W_t \,,$$

though we computed the optimal controller with the matrix B. Define an error signal, $E_t = X_t - \widehat{x}_t$, that measures the current deviation between the actual state and the estimate. Then, using the fact that $u_t = -K\widehat{x}_t$, we get the closed loop dynamics

$$\begin{bmatrix} \widehat{X}_{t+1} \\ E_{t+1} \end{bmatrix} = \begin{bmatrix} A - BK & LC \\ (B - B_\star)K & A - LC \end{bmatrix} \begin{bmatrix} \widehat{X}_t \\ E_t \end{bmatrix} + \begin{bmatrix} LV_t \\ W_t - LV_t \end{bmatrix} \,.$$

When $B = B_\star$, the bottom left block is equal to zero. The system is then stable provided $A - BK$ and $A - LC$ are both stable matrices (i.e., have eigenvalues with magnitude less than 1). However, small perturbations in the off-diagonal block can make the matrix unstable. For intuition, consider the matrix

$$\begin{bmatrix} 0.9 & 1 \\ 0 & 0.8 \end{bmatrix} \,.$$

The eigenvalues of this matrix are 0.9 and 0.8, so the matrix is clearly stable. But the matrix

$$\begin{bmatrix} 0.9 & 1 \\ t & 0.8 \end{bmatrix}$$

has an eigenvalue greater than 0 if $t > 0.02$. So a tiny perturbation significantly shifts the eigenvalues and makes the matrix unstable.

Similar things happen in LQG. Let's return to our simple dynamics inspired by Newton's Laws of Motion

$$A = \begin{bmatrix} 1 & 1 \\ 0 & 1 \end{bmatrix} , \quad B = \begin{bmatrix} 0 \\ 1 \end{bmatrix} , \quad C = \begin{bmatrix} 1 & 0 \end{bmatrix} \,.$$

And let's use *any* cost matrices Φ and Ψ. We assume that the noise variances are

$$\mathbb{E}\left[W_t W_t^\top\right] = \begin{bmatrix} 1 & 2 \\ 2 & 4 \end{bmatrix} , \quad \mathbb{E}\left[V_t^2\right] = \sigma^2 \,.$$

The open loop system here is unstable, having two eigenvalues at 1. We can stabilize the system only by modifying the second state. The state disturbance is aligned along the direction of the vector $[1/2; 1]$, and the state cost only penalizes states aligned with this disturbance. The SDM goal is simply to remove as much signal as possible in the $[1; 1]$ direction without using large inputs. We only are able to measure the first component of the state, and this measurement is corrupted by Gaussian noise.

What does the optimal policy look like? Perhaps unsurprisingly, it focuses all of its energy on ensuring that there is little state signal along the disturbance direction. The optimal L matrix is

$$L = \begin{bmatrix} 3 - d_1 \\ 2 - d_2 \end{bmatrix},$$

where d_1 and d_2 are small positive numbers that go to 0 as σ goes to 0. The optimal K will have positive coefficients whenever we choose for Φ and Ψ to be positive semidefinite: if K has a negative entry, it will necessarily not stabilize (A, B).

Now what happens when we have model mismatch? Let's assume for simplicity that $\sigma = 0$. If we set $B_\star = tB$ and use the formula for the closed loop above, we see that closed loop state transition matrix is

$$A_{cl} = \begin{bmatrix} 1 & 1 & 3 & 0 \\ -k_1 & 1 - k_2 & 2 & 0 \\ 0 & 0 & -2 & 1 \\ k_1(1-t) & k_2(1-t) & -2 & 1 \end{bmatrix}.$$

It's straightforward to check that when $t = 1$ (i.e., no model mismatch), the eigenvalues of $A - BK$ and $A - LC$ all have real parts with magnitude less than or equal to 1. For the full closed loop matrix, analytically computing the eigenvalues themselves is a pain, but we can prove instability by looking at a characteristic polynomial. For a matrix to have all of its eigenvalues in the left half plane, its characteristic polynomial necessarily must have all positive coefficients. If we look at the linear term in the characteristic polynomial of $-I - A_{cl}$, we see that if $t > 1$, A_{cl} must have an eigenvalue with real part less than -1, and hence the closed loop is unstable. This is a very conservative condition, and we could get a tighter bound if we'd like, but it's good enough to reveal some paradoxical properties of LQG. The most striking is that if we build a sensor that gives us a better and better measurement, our system becomes more and more fragile to perturbation and model mismatch. For machine learning scientists, this seems to go against all of our training. How can a system become *less* robust if we improve our sensing and estimation?

Let's look at the example in more detail to get some intuition for what's happening. When the sensor noise gets small, the optimal Kalman Filter is more aggressive. The filter rapidly damps any errors in the disturbance direction $[1; 1/2]$ and, as σ decreases, it damps the $[1; 1]$ direction less. When $t \neq 1$, $B - B_\star$ is aligned in the $[0; 1]$ and can be treated as a disturbance signal. This undamped component of the error is fed errors from the state estimate, and these errors compound each other. Since we spend so much time focusing on our control along the direction of the injected state noise, we become highly susceptible to errors in a different direction and these are the exact errors that occur when there is a gain mismatch between the model and reality.

The fragility of LQG has many takeaways. It highlights that noiseless state measurement can be a dangerous modeling assumption, because it is then optimal to trust our model too much. Model mismatch must be explicitly accounted for when designing the decision making policies.

This should be a cautionary tale for modern AI systems. Most papers in reinforcement learning consider MDPs where we perfectly measure the system state. Building an entire field around optimal actions with perfect state observation builds too much optimism. Any realistic scenario is going to have partial state observation, and such problems are much thornier.

A second lesson is that it is not enough to just improve the prediction components in feedback systems that are powered by machine learning. Improving prediction will increase sensitivity to modeling errors in some other part of the engineering pipeline, and these must all be accounted for together to ensure safe and successful decision making.

Chapter notes

This chapter and the previous chapter overlap significantly with a survey of reinforcement learning by Recht, which contains additional connections to continuous control.[218]

Bertsekas has written several valuable texts on reinforcement learning from different perspectives. His seminal book with Tsitsiklis established the mathematical formalisms of Neurodynamic Programming that most resemble contemporary reinforcement learning.[238] The second volume of his Dynamic Programming Book covers many of the advanced topics in approximate dynamic programming and infinite horizon dynamic programming.[248] And his recent book on reinforcement learning builds ties with his earlier work and recent advances in reinforcement learning post AlphaGo.[249]

For more on bandits from a theoretical perspective, the reader is invited to consult the comprehensive book by Lattimore and Szepesvari.[250] Agarwal et al. provide a thorough introduction to the theoretical aspects of reinforcement learning from the perspective of learning theory.[241] The control-theoretic perspective on reinforcement learning is called *dual control*. Its originated at a similar time to that of reinforcement learning, and many attribute Feldbaum's work as the origin point.[251] Wittenmark surveys the history of this topic, its limitations, and its comparison to certainty equivalence methods.[252] To further explore the limits of classical optimal control and how to think about robustness, Stein's "Respect the unstable" remains a classic lecture on the subject.[253]

Chapter 13

Epilogue

Unknown outcomes often follow patterns found in past observations. But when do they not? As powerful as statistical patterns are, they are not without limitations. Every discipline built on the empirical law also experiences its failure.

In fact, Halley's contemporaries already bore witness. Seeking to increase revenue still, despite the sale of life annuities, King William III desired to tax his citizens in proportion to their wealth. An income tax appeared too controversial and unpopular with his constituents, so the king's advisors had to come up with something else. In 1696, the king introduced a property tax based on the number of windows in a house. It stands to reason that the wealth of a family correlated strongly with the number of windows in their home. So, the window tax looked quite reasonable from a statistical perspective.

Although successful on the whole and adopted by many other countries, the window tax had a peculiar side effect. People adjusted. Increasingly, houses would have bricked-up window spaces. In Edinburgh an entire row of houses featured no bedroom windows at all. The correlation between the number of windows and wealth thus deteriorated.

The problem with the window tax foretold a robust limitation of prediction. Datasets display a static snapshot of a population. Predictions on the basis of data are accurate only under an unspoken stability assumption. Future observations must follow the same data generating process. It's the "more of the same" principle that we call generalization in supervised learning.

However, predictions often motivate consequential actions in the real world that change the populations we observe. Chemist and technology critic Ursula Franklin summarizes the problem aptly in her 1989 book called *The Real World of Technology*:

> [T]echnologies are developed and used within a particular social, economic, and political context. They arise out of a social structure,

> they are grafted on to it, and they may reinforce it or destroy it, often in ways that are neither foreseen nor foreseeable.[254]

Franklin continues:

> [C]ontext is not a passive medium but a dynamic counterpart. The responses of people, individually, and collectively, and the responses of nature are often underrated in the formulation of plans and predictions.

Franklin understood that predictions are not made in a vacuum. They are agents of change through the actions they prompt. Decisions are always part of an evolving environment. It's this dynamic environment that determines the merit of a decision.

Predictions can fail catastrophically when the underlying population is subject to unmodeled changes. Even benign changes to a population, sometimes called distribution shift, can sharply degrade the utility of statistical models. Numerous results in machine learning are testament to the fragility of even the best performing models under changing environments.

Other disciplines have run into the same problem. In his influential critique from 1976, economist Robert Lucas argued that patterns found in historical macroeconomic data are an inadequate basis of policy making, since any policy would inevitably perturb those statistical patterns. Subsequently, economists sought to ground macroeconomics in the microeconomic principles of utility theory and rational behavior of the individual, an intellectual program known as microfoundations dominant to this day. The hope was that microfoundations would furnish a more reliable basis of economic policy making.

It is tempting to see dynamic modeling as a possible remedy to the problem Lucas describes. However, Lucas' critique *was* about dynamic models. Macroeconomists at the time were well aware of dynamic programming and optimal control. A survey of control-theoretic tools in macroeconomics from 1976 starts with the lines:

> In the past decade, a number of engineers and economists have asked the question: "If modern control theory can improve the guidance of airplanes and spacecraft, can it also help in the control of inflation and unemployment?"[255]

If anything, the 1960s and 1970s were the heyday of dynamic modeling. Entire disciplines, such as *system dynamics*, attempted to create dynamic models of complex social systems, such as corporations, cities, and even the Western industrial world. Proponents of system dynamics used simulations of these models to motivate consequential policy propositions. Reflecting on these times, economist Schumacher wrote in 1973:

> There have never been so many futurologists, planners, forecasters, and model-builders as there are today, and the most intriguing product of technological progress, the computer, seems to offer untold new possibilities. ... Are not such machines just what we have been waiting for?[256]

It was not the lack of dynamic models that Lucas criticized, it was the fact that policy may invalidate the empirical basis of the model. Lucas' critique puts pressure how we come to *know* a model. Taking action can invalidate not just a particular model but also disrupt the social and empirical facts from which we derived the model.

If economics reckoned with this problem decades ago, it's worth taking a look at how the field has developed since. Oversimplifying greatly, the ambitious macroeconomic theorizing of the twentieth century gave way to a greater focus on microeconomics and empirical work. Field experiments and causal inference, in particular, are now at the forefront of economic research.

Fundamental limitations of dynamic models not only surfaced in economics, they were also called out by control theorists themselves. In a widely heralded plenary lecture at the 1989 IEEE Conference on Decision and Control, Gunter Stein argued against "the increasing worship of abstract mathematical results in control at the expense of more specific examinations of their practical, physical consequences." Stein warned that mathematical and algorithmic formulations often elided fundamental physical limitations and trade-offs that could lead to catastrophic consequences.

Unstable systems illustrate this point. A stable system has the property that no matter how you disturb the system, it will always come back to rest. If you heat water on the stove, it will always eventually return to room temperature. An unstable system on the other hand can evolve away from a natural equilibrium exponentially quickly, like a contagious pathogen. From a computational perspective, however, there is no more difficulty in mathematically solving a sequential decision making problem with unstable dynamics than in solving one with stable dynamics. We can write down and solve decision making problems in both cases, and they appear to be of equal computational difficulty. But in reality, unstable systems are dangerous in a way that stable systems are not. Small errors get rapidly amplified, possibly resulting in catastrophe. Likely the most famous such catastrophe is the Chernobyl disaster, which Stein described as the failure to "respect the unstable" inherent in the reactor design.

As the artificial intelligence and machine learning communities increasingly embrace dynamic modeling, they will inevitably relearn these cautionary lessons of days past.

Beyond pattern classification?

Part of the recent enthusiasm for causality and reinforcement learning stems from the hope that these formalisms might address some of the inherent issues with the static pattern classification paradigm. Indeed, they might. But neither causality nor reinforcement learning are a panacea. Without hard-earned substantive domain knowledge to guide modeling and mathematical assumptions, there is little that sets these formalisms apart from pattern classification. The reliance on subject matter knowledge stands in contrast with the nature of recent advances in machine learning that largely did without such—and that was the point.

Looking ahead, the space of machine learning beyond pattern classification is full of uncharted territory. In fact, even the basic premise that there is such a space is not entirely settled.

Some argue that as a practical matter machine learning will proceed in its current form. Those who think so would see progress coming from faster hardware, larger datasets, better benchmarks, and increasingly clever ways of reducing new problems to pattern classification. This position isn't unreasonable in light of historical or recent developments. Pattern classification has reemerged several times over the past 70 years, and each time it has shown increasingly impressive capabilities.

We can try to imagine what replaces pattern recognition when it falls out of favor. And perhaps we can find some inspiration by returning one last time to Edmund Halley. Halley is more well known for astronomy than for his life table. Much of astronomy before the seventeenth century was more similar to pattern recognition than fundamental physics. Halley himself had used curve-fitting methods to predict the paths of comets, but found notable errors in his predictions for the comet Kirch. He discussed his calculations with Isaac Newton, who solved the problem by establishing a fundamental description of the laws of gravity and motion. Halley, so excited by these results, paid to publish Newton's magnum opus *Philosophiæ Naturalis Principia Mathematica*.

Even if it may not be physics once again or on its own, similarly disruptive conceptual departures from pattern recognition may be viable and necessary for machine learning to become a safe and reliable technology in our lives.

We hope that our story about machine learning was helpful to those who aspire to write its next chapters.

Chapter 14

Mathematical Background

The main mathematical tools of machine learning are optimization and statistics. At their core are concepts from multivariate calculus and probability. Here, we briefly review some of the concepts from calculus and probability that we will frequently make use of in the book.

Common notation

- Lowercase letters u, v, w, x, y, z typically denote vectors. We use both $\langle u, v \rangle$ and $u^T v$ to denote the inner product between vectors u and v.
- Capital letters X, Y, Z typically denote random variables.
- $\mathbb{P}[A \mid B]$ denotes the conditional probability of an event A conditional on an event B
- The gradient $\nabla f(x)$ of a function $f \colon \mathbb{R}^d \to \mathbb{R}$ at a point $x \in \mathbb{R}^d$ refers to the vector of partial derivatives of f evaluated at x.
- Identity matrix is I.
- The first k positive integers are $[k] = \{1, 2, \ldots, k\}$.

Multivariable calculus and linear algebra

Positive definite matrices

Positive definite matrices are central to both optimization algorithms and statistics. In this section, we quickly review some of the core properties that we will use throughout the book.

A matrix M is *positive definite* (pd) if it is symmetric $M = M^T$ and $z^T M z > 0$ for all nonzero $z \in \mathbb{R}^d$. We denote this as $M \succ 0$. A matrix M is *positive semidefinite* (psd) if it is symmetric and $z^T M z \geq 0$ for all nonzero z. We denote this as $M \succeq 0$. All pd matrices are psd, but not vice versa.

Some of the main properties of positive semidefinite matrices include:

1. If $M_1 \succeq 0$, and $M_2 \succeq 0$, then $M_1 + M_2 \succeq 0$.
2. $a \in \mathbb{R}$, $a \geq 0$ implies $aM \succeq 0$.
3. For any matrix F, FF^T and $F^T F$ are both psd. Conversely, if M is psd there exists an F such that $M = FF^T$.

Note that (1) and (2) still hold if "psd" is replaced with "pd."" That is, the sum of two pd matrices is pd. And multiplying a pd matrix by a positive scalar preserves positive definiteness.

Recall that λ is a eigenvalue of a square matrix M if there exists a nonzero $x \in \mathbb{R}^d$ such that $Mx = \lambda x$. Eigenvalues of psd matrices are all nonnegative. Eigenvalues of pd matrices are all positive. This follows by multiplying the equation $Ax = \lambda x$ on the left by x^T.

Gradients, Taylor's Theorem, and infinitesimal approximation

Let $\Phi : \mathbb{R}^d \to \mathbb{R}$. Recall from multivariable calculus that the *gradient* of Φ at a point w is the vector of partial derivatives

$$\nabla \Phi(w) = \begin{bmatrix} \frac{\partial \Phi(w)}{\partial x_1} \\ \frac{\partial \Phi(w)}{\partial x_2} \\ \vdots \\ \frac{\partial \Phi(w)}{\partial x_d} \end{bmatrix}.$$

Sometimes we write $\nabla_x \Phi(w)$ to make clear which functional argument we are referring to.

One of the most important theorems in calculus is *Taylor's theorem*, which allows us to approximate smooth functions by simple polynomials. The following simplified version of Taylor's Theorem is used throughout optimization. This form of Taylor's theorem is sometimes called the multivariable mean-value theorem. We will use this at multiple points to analyze algorithms and understand the local properties of functions.

Theorem 13. ***Taylor's Theorem.***

- *If Φ is continuously differentiable, then, for some $t \in [0, 1]$,*

$$\Phi(w) = \Phi(w_0) + \nabla \Phi(tw + (1-t)w_0)^T (w - w_0).$$

- *If Φ is twice continuously differentiable, then*

$$\nabla \Phi(w) = \nabla \Phi(w_0) + \int_0^1 \nabla^2 \Phi(tw + (1-t)w_0)(w - w_0) \mathrm{d}t$$

and, for some $t \in [0,1]$,

$$\Phi(w) = \Phi(w_0) + \nabla\Phi(w_0)^T(w - w_0) + \frac{1}{2}(w - w_0)^T \nabla^2\Phi(tw + (1-t)w_0)^T(w - w_0) .$$

Taylor's theorem can be used to understand the local properties of functions. For example,

$$\Phi(w + \epsilon v) = \Phi(w) + \epsilon\nabla\Phi(w)^T v + \frac{\epsilon^2}{2} v^T \nabla^2\Phi(w + \delta v)^T v$$

for some $0 \le \delta \le \epsilon$. This expression states that

$$\Phi(w + \epsilon v) = \Phi(w) + \epsilon\nabla\Phi(w)^T v + \Theta(\epsilon^2) ,$$

so, to first order, we can approximate Φ by a linear function.

Jacobians and the multivariate chain rule

The matrix of first-order partial derivatives of a multivariate mapping $\Phi\colon \mathbb{R}^n \to \mathbb{R}^m$ is called *Jacobian matrix*. We define the Jacobian of Φ with respect to a variable x evaluated at a value w as the $m \times n$ matrix

$$\mathsf{D}_x\Phi(w) = \left[\frac{\partial\Phi_i(w)}{\partial x_j}\right]_{i=1\ldots m, j=1\ldots n} .$$

The ith row of the Jacobian therefore corresponds to the transpose of the familiar gradient $\nabla_x^T\Phi_i(w)$ of the ith coordinate of Φ. In particular, when $m = 1$ the Jacobian corresponds to the transpose of the gradient.

The first-order approximation given by Taylor's theorem directly extends to multivariate functions via the Jacobian matrix. So does the *chain rule* from calculus for computing the derivatives of function compositions.

Let $\Phi\colon \mathbb{R}^n \to \mathbb{R}^m$ and $\Psi\colon \mathbb{R}^m \to \mathbb{R}^k$. Then, we have

$$\mathsf{D}_x\Psi \circ \Phi(w) = \mathsf{D}_{\Phi(w)}\Psi(\Phi(w))\mathsf{D}_x\Phi(w) .$$

As we did with the gradient notation, when the variable x is clear from context we may drop it from our notation and write $\mathsf{D}\Phi(w)$.

Probability

Contemporary machine learning uses probability as its primary means of quantifying uncertainty. Here we review some of the basics we will make use of in this course. This will also allow us to fix notation.

We note that oftentimes, mathematical rigor gets in the way of explaining concepts. So we will attempt to only introduce mathematical machinery when absolutely necessary.

Probability is a function on sets. Let $\mathcal{X}$ denote the sample set. For every $A \subset \mathcal{X}$, we have

$$0 \leq \mathbb{P}[A] \leq 1, \qquad \mathbb{P}[\mathcal{X}] = 1, \qquad \mathbb{P}[\emptyset] = 0,$$

and

$$\mathbb{P}[A \cup B] + \mathbb{P}[A \cap B] = \mathbb{P}[A] + \mathbb{P}[B].$$

This implies that

$$\mathbb{P}[A \cup B] = \mathbb{P}[A] + \mathbb{P}[B]$$

if and only if $\mathbb{P}[A \cap B] = 0$. We always have the inequality

$$\mathbb{P}[A \cup B] \leq \mathbb{P}[A] + \mathbb{P}[B].$$

By induction, we get the union bound

$$\mathbb{P}\left[\bigcup_i A_i\right] \leq \sum_i \mathbb{P}[A_i].$$

Random variables and vectors

Random variables are a particular way of characterizing outcomes of random processes. We will use capital letters like X, Y, and Z to denote such random variables. The sample space of a random variable will be the set where a variable can take values. Events are simply subsets of possible values. Common examples we will encounter in this book are

- **Probability that a random variable has a particular value**. This will be denoted as $\mathbb{P}[X = x]$. Note here that we use a lowercase letter to denote the value that the random variable might take.
- **Probability that a random variable satisfies some inequality**. For example, the probability that X is less than a scalar t will be denoted as $\mathbb{P}[X \leq t]$.

A *random vector* is a random variable whose sample space consists of R^d. We will not use notation to distinguish between vectors and scalars in this text.

Densities

Random vectors are often characterized by *probability densities* rather than by probabilities. The density p of a random variable X is defined by its relation to probabilities of sets:

$$\mathbb{P}[X \in A] = \int_{x \in A} p(x)\mathrm{d}x.$$

Expectations

If f is a function on R^d and X is a random vector, then the expectation of f is given by

$$\mathbb{E}[f(X)] = \int f(x)p(x)\mathrm{d}x\,.$$

If A is a set, the *indicator function of the set* is the function

$$I_A(x) = \begin{cases} 1 & \text{if } x \in A \\ 0 & \text{otherwise}\,. \end{cases}$$

Note that the expectation of an indicator function is a probability:

$$\mathbb{E}[I_A(X)] = \int_{x\in A} p(x)\mathrm{d}x = \mathbb{P}[X \in A]\,.$$

This expression links the three concepts of expectation, density, and probability together.

Note that the expectation operator is linear:

$$\mathbb{E}[af(X) + bg(X)] = a\,\mathbb{E}[f(X)] + b\,\mathbb{E}[g(x)]\,.$$

Two other important expectations are the mean and covariance. The *mean* of a random variable is the expected value of the identity function:

$$\mu_X := \mathbb{E}[X] = \int xp(x)\mathrm{d}x\,.$$

The *covariance* of a random variable is the matrix

$$\Sigma_X := \mathbb{E}[(X - \mu_X)(X - \mu_X)^T]\,.$$

Note that covariance matrices are positive semidefinite. To see this, take a nonzero vector z and compute

$$z^T\Sigma_X z := \mathbb{E}[z^T(X - \mu_X)(X - \mu_X)^T z] = \mathbb{E}[((X - \mu_X)^T z)^2]\,.$$

Since the term inside the expectation is nonnegative, the expectation is nonnegative as well.

Important examples of probability distributions

- **Bernoulli random variables.** A Bernoulli random variable X can take two values, 0 and 1. In such a case $\mathbb{P}[X = 1] = 1 - \mathbb{P}[X = 0]$.

- **Gaussian random vectors.** Gaussian random vectors are the most ubiquitous real-valued random vectors. Their densities are parameterized only by their mean and covariance:

$$p(x) = \frac{1}{\det(2\pi\Sigma)^{1/2}} \exp\left(-\tfrac{1}{2}(x-\mu_X)^T\Sigma^{-1}(x-\mu_X)\right) .$$

  Gaussian random variables are often called "normal" random variables. We denote the distribution of a normal random variable with mean μ and covariance Σ as

$$\mathcal{N}(\mu,\Sigma) .$$

  The reason Gaussian random variables are ubiquitous is because of the central limit theorem: averages of many independent random variables tend to look like Gaussian random variables.

Conditional probability and Bayes' Rule

Conditional probability is applied quite cavalierly in machine learning. It's actually very delicate and should only be applied when we really know what we're doing.

$$\mathbb{P}[A|B] = \frac{\mathbb{P}[A \cap B]}{\mathbb{P}[B]} .$$

A and B are said to be *independent* if $\mathbb{P}[A|B] = \mathbb{P}[A]$. Note that from the definition of conditional probability, A and B are independent if and only if

$$\mathbb{P}[A \cap B] = \mathbb{P}[A]\,\mathbb{P}[B] .$$

Bayes' rule is an immediate corollary of the definition of conditional probability. In some sense, it's just a restatement of the definition:

$$\mathbb{P}[A|B] = \frac{\mathbb{P}[B|A]\,\mathbb{P}[A]}{\mathbb{P}[B]}$$

This is commonly applied when A is one of a set of several alternatives. Suppose A_i are a collection of disjoint sets such that $\cup_i A_i = \mathcal{X}$, then for each i, Bayes' Rule states

$$\mathbb{P}[A_i|B] = \frac{\mathbb{P}[B|A_i]\,\mathbb{P}[A_i]}{\sum_j \mathbb{P}[B|A_j]\,\mathbb{P}[A_j]} .$$

This shows that if we have models of the likelihood of B under each alternative A_i and if we have beliefs about the probability of each A_i, we can compute the probability of observing A_i under the condition that B has occurred.

Conditional densities

Suppose X and Z are random variables whose joint distribution is continuous. If we try to write down the conditional distribution for X given $Z = z$, we find

$$\mathbb{P}[X \in A | Z = z] = \frac{\mathbb{P}[X \in A \cap Z = z]}{\mathbb{P}[Z = z]}.$$

Both the numerator and denominator are equal to zero. In order to have a useful formula, we can appeal to densities:

$$\begin{aligned}
\mathbb{P}[x \in A | z \leq Z \leq z + \epsilon] &= \frac{\int_z^{z+\epsilon} \int_{x \in A} p(x, z') \mathrm{d}x \mathrm{d}z'}{\int_z^{z+\epsilon} p(z') \mathrm{d}z'} \\
&\approx \frac{\epsilon \int_{x \in A} p(x, z)}{\epsilon p(z) \mathrm{d}z} \\
&= \int_{x \in A} \frac{p(x, z)}{p(z)} \mathrm{d}x.
\end{aligned}$$

Letting ϵ go to zero, this calculation shows that we can use the *conditional density* to compute the conditional probabilities of X when $Z = z$:

$$p(x|z) := \frac{p(x, z)}{p(z)}.$$

Conditional expectation and the law of iterated expectation

Conditional expectation is shorthand for computing expected values with respect to conditional probabilities:

$$\mathbb{E}[f(x, z) | Z = z] = \int f(x, z) p(x|z) \mathrm{d}x.$$

An important formula is the law of iterated expectation:

$$\mathbb{E}[f(x, z)] = \mathbb{E}[\mathbb{E}[f(x, z) | Z = z]]$$

This formula follows because

$$\begin{aligned}
\mathbb{E}[f(x, z)] &= \int \int f(x, z) p(x, z) \mathrm{d}x \mathrm{d}z \\
&= \int \int f(x, z) p(x|z) p(z) \mathrm{d}x \mathrm{d}z \\
&= \int \left(\int f(x, z) p(x|z) \mathrm{d}x \right) p(z) \mathrm{d}z.
\end{aligned}$$

Estimation

This book devotes much of its attention to probabilistic decision making. A different but related statistical problem is *parameter estimation*. Assuming that data X is generated by a statistical model, we'd like to infer some *nonrandom* property about its distribution. The most canonical example here would be estimating the mean or variance of the distribution. Note that estimating these parameters has a different flavor than decision theory. In particular, our framework of risk minimization no longer applies.

If we aim to minimize a functional

$$\text{minimize}_f \; \mathbb{E}[loss(\vartheta, f(x))]$$

then the optimal choice is to set $f(x) = \vartheta$. But we don't know this parameter in the first place. So we end up with an algorithm that's not implementable.

Instead, what we do in estimation theory is pose a variety of plausible estimators that might work for a particular parameter and consider the efficacy of these parameters in different settings. In particular, we'd like estimators that take a set of observations $S = (x_1, \ldots, x_n)$ and return a guess for the parameter whose value improves as n increases:

$$\lim_{n\to\infty} \mathbb{E}_S[loss(\vartheta, \widehat{\vartheta}(S))] = 0\,.$$

Even though estimators are constructed from data, their design and implementation require a good deal of knowledge about the underlying probability distribution. Because of this, estimation is typically considered to be part of classical statistics and not machine learning. Estimation theory has a variety of powerful tools that are aimed at producing high-quality estimators, and is certainly worth learning more about. We need rudimentary elements of estimation to understand popular baselines and algorithms in causal inference and reinforcement learning.

Plug-in Estimators

We will restrict our attention to *plug-in estimators*. Plug-in estimators are functions of the moments of probability distributions. They are plug-in because we replace the true distribution with the empirical distribution. To be precise, suppose there exist vector valued functions g and ψ such that $\vartheta = g(\mathbb{E}[\psi(x)])$. Then, given a dataset, $S = (x_1, \ldots, x_n)$, the associated plug-in estimator of ϑ is

$$\widehat{\vartheta}(S) = g\left(\frac{1}{n}\sum_{i=1}^{n}\psi(x_i)\right)\,,$$

that is, we replace the expectation with the sample average. There are canonical examples of plug-in estimators.

1. *The sample mean.* The sample mean is the plug-in estimator where g and ψ are both the identity functions.

2. *The sample covariance.* The sample covariance is

$$\widehat{\Sigma}_x = \sum_{i=1}^n x_i x_i^T - \left(\frac{1}{n}\sum_{i=1}^n x_i\right)\left(\sum_{i=1}^n x_i\right)^T .$$

   From this formula, we can take

$$\psi(x) = \begin{bmatrix} 1 \\ x \end{bmatrix} \begin{bmatrix} 1 \\ x \end{bmatrix}^T \quad \text{and} \quad g\left(\begin{bmatrix} A & B \\ B^T & C \end{bmatrix}\right) = C - BB^T .$$

3. *Least squares estimator.* Suppose we have three random vectors, y, x, and v and we assume that v and x are zero-mean and uncorrelated and that $y = Ax + v$ for some matrix A. Let's suppose we'd like to estimate A from a set of pairs $S = ((x_1, y_1), \ldots, (x_n, y_n))$. One can check that

$$A = \Sigma_{yx}\Sigma_x^{-1} .$$

   Hence, the plug-in estimator would use the sample covariances:

$$\widehat{A} = \left(\sum_{i=1}^n y_i x_i^T\right)\left(\sum_{i=1}^n x_i x_i^T\right)^{-1} .$$

   In this case, we have the formulation

$$\psi(x) = \begin{bmatrix} x \\ y \end{bmatrix} \begin{bmatrix} x \\ y \end{bmatrix}^T \quad \text{and} \quad g\left(\begin{bmatrix} A & B \\ B^T & C \end{bmatrix}\right) = BA^{-1} .$$

Convergence rates

In our study of generalization, we reasoned that the empirical risk should be close to the true risk because sample averages should be close to population values. A similar reasoning holds true for plug-in estimators: smooth functions of sample averages should be close to their population counterparts.

We covered the case of the sample mean in our discussion of generalization. To recall, suppose x is a Bernoulli random variable with mean p. Let $x_1, \ldots, x_n$ be independent and identically distributed as x. Then Hoeffding's inequality states that

$$\mathbb{P}\left[\left|\frac{1}{n}\sum_{i=1}^n x_i - p\right| > \epsilon\right] \leq 2\exp(-2n\epsilon^2) .$$

Or, in other words, with probability $1 - \delta$,

$$\left| \frac{1}{n} \sum_{i=1}^{n} x_i - p \right| \leq \sqrt{\frac{\log(2/\delta)}{2n}} .$$

Let's consider a simple least squares estimator. Suppose we know that $y = w^T x + v$ where w and x are vectors, w is deterministic, and x and v are uncorrelated. Consider the least squares estimator $\widehat{w}_S$ from n data points. The estimation error in w is the vector $e_S = \widehat{w}_S - w$. The expectation of e_S is zero and the expected norm of the error is given by

$$\mathbb{E}\left[\|e_S\|^2\right] = \text{Trace}\left(\left(\sum_{i=1}^{n} x_i x_i^T\right)^{-1}\right) .$$

This error is small if the sample covariance has large eigenvalues. Indeed, if λ_S denotes the minimum eigenvalue of the sample covariance of x, then

$$\mathbb{E}\left[\|e_S\|^2\right] \leq \frac{d}{n}\lambda_S .$$

This expression suggests that the distribution of x must have density that covers all directions somewhat equally in order for the least squares estimator to have good performance. On top of this, we see that the squared error decreases roughly as d/n. Hence, we need far more measurements than dimensions to find a good estimate of w. This is in contrast to what we studied in classification. Most of the generalization bounds for classification we derived were *dimension-free* and only depended on properties like the margin of the data. In contrast, in parameter estimation, we tend to get results that scale as number of parameters over number of data points. This rough rule of thumb that the error scales as the ratio of number of parameters to number of data points tends to be a good guiding principle when attempting to understand convergence rates of estimators.

Bibliography

[1] New navy device learns by doing; psychologist shows embryo of computer designed to read and grow wiser. *The New York Times*, 1958.

[2] David R. Bellhouse. A new look at Halley's life table. *Journal of the Royal Statistical Society: Series A (Statistics in Society)*, 174(3):823–832, 2011.

[3] James E. Ciecka. Edmond Halley's life table and its uses. *Journal of Legal Economics*, 15:65–74, 2008.

[4] Karl Pearson and Egon S. Pearson. The history of statistics in the 17th and 18th centuries against the changing background of intellectual, scientific and religious thought. *British Journal for the Philosophy of Science*, 32(2):177–183, 1981.

[5] Ian Hacking. *The Emergence of Probability: A Philosophical Study of Early Ideas about Probability, Induction and Statistical Inference*. Cambridge University Press, 2006.

[6] David A. Mindell. *Between Human and Machine: Feedback, Control, and Computing before Cybernetics*. JHU Press, 2002.

[7] Ronald R. Kline. *The Cybernetics Moment: Or Why We Call Our Age the Information Age*. JHU Press, 2015.

[8] Steve J. Heims. *The Cybernetics Group*. MIT Press, 1991.

[9] James Beniger. *The control revolution: Technological and economic origins of the information society*. Harvard University Press, 1986.

[10] Ryszard S. Michalski, Jamie G. Carbonell, and Tom M. Mitchell, editors. *Machine Learning: An Artificial Intelligence Approach*. Springer, 1983.

[11] Pat Langley. The changing science of machine learning, 2011.

[12] Mark Liberman. Obituary: Fred Jelinek. *Computational Linguistics*, 36(4):595–599, 2010.

[13] Kenneth Ward Church. Emerging trends: A tribute to Charles Wayne. *Natural Language Engineering*, 24(1):155–160, 2018.

[14] Mark Liberman and Charles Wayne. Human language technology. *AI Magazine*, 41(2):22–35, 2020.

[15] James L. McClelland, David E. Rumelhart, and PDP Research Group. Parallel distributed processing. *Explorations in the Microstructure of Cognition*, 2:216–271, 1986.

[16] Michael I. Jordan. Artificial intelligence—the revolution hasn't happened yet. *Harvard Data Science Review*, 1(1), 2019.

[17] Ruha Benjamin. *Race after Technology*. Polity, 2019.

[18] Ben Hutchinson and Margaret Mitchell. 50 years of test (un) fairness: Lessons for machine learning. In *Conference on Fairness, Accountability, and Transparency*, pages 49–58, 2019.

[19] Solon Barocas, Moritz Hardt, and Arvind Narayanan. *Fairness and Machine Learning*. fairmlbook.org, 2019. `http://www.fairmlbook.org`.

[20] Jon M. Kleinberg, Sendhil Mullainathan, and Manish Raghavan. Inherent trade-offs in the fair determination of risk scores. In *Innovations in Theoretical Computer Science*, 2017.

[21] Alexandra Chouldechova. Fair prediction with disparate impact: A study of bias in recidivism prediction instruments. *Big Data*, 5(2):153–163, 2017.

[22] Julia Angwin, Jeff Larson, Surya Mattu, and Lauren Kirchner. Machine bias. *ProPublica*, May 2016.

[23] William Dieterich, Christina Mendoza, and Tim Brennan. COMPAS risk scales: Demonstrating accuracy equity and predictive parity. Technical report, 2016.

[24] Jerzy Neyman and Egon S. Pearson. On the use and interpretation of certain test criteria for purposes of statistical inference: Part I. *Biometrika*, pages 175–240, 1928.

[25] Jerzy Neyman and Egon S. Pearson. On the problem of the most efficient tests of statistical hypotheses. *Philosophical Transactions of the Royal Society of London. Series A*, 231(694-706):289–337, 1933.

[26] Abraham Wald. Contributions to the theory of statistical estimation and testing hypotheses. *The Annals of Mathematical Statistics*, 10(4):299–326, 1939.

[27] Dimitri P. Bertsekas and John N. Tsitsiklis. *Introduction to Probability*. Athena Scientific, 2nd edition, 2008.

[28] W. Wesley Peterson, Theodore G. Birdsall, and William C. Fox. The theory of signal detectability. *Transactions of the IRE*, 4(4):171–212, 1954.

[29] Wilson P. Tanner Jr. and John A. Swets. A decision-making theory of visual detection. *Psychological Review*, 61(6):401, 1954.

[30] Chao Kong Chow. An optimum character recognition system using decision functions. *IRE Transactions on Electronic Computers*, (4):247–254, 1957.

[31] Wilbur H. Highleyman. Linear decision functions, with application to pattern recognition. *Proceedings of the IRE*, 50(6):1501–1514, 1962.

[32] Meredith Broussard. *Artificial Unintelligence: How Computers Misunderstand the World*. MIT Press, 2018.

[33] Virginia Eubanks. *Automating Inequality: How High-Tech Tools Profile, Police, and Punish the Poor*. St. Martin's Press, 2018.

[34] Safiya Umoja Noble. *Algorithms of Oppression: How Search Engines Reinforce Racism*. NYU Press, 2018.

[35] Cathy O'Neil. *Weapons of Math Destruction: How Big Data Increases Inequality and Threatens Democracy*. Broadway Books, 2016.

[36] Frank Rosenblatt. The perceptron: A probabilistic model for information storage and organization in the brain. *Psychological Review*, pages 65–386, 1958.

[37] Michael J. Kearns, Robert E. Schapire, and Linda M. Sellie. Toward efficient agnostic learning. *Machine Learning*, 17(2-3):115–141, 1994.

[38] Albert B. J. Novikoff. On convergence proofs on perceptrons. In *Symposium on the Mathematical Theory of Automata*, pages 615–622, 1962.

[39] Vladimir Vapnik and Alexey Chervonenkis. *Theory of Pattern Recognition: Statistical Learning Problems*. Nauka, 1974. In Russian.

[40] Frank Rosenblatt. *Two Theorems of Statistical Separability in the Perceptron*. United States Department of Commerce, 1958.

[41] Frank Rosenblatt. *Principles of Neurodynamics: Perceptions and the Theory of Brain Mechanisms*. Spartan, 1962.

[42] Hans-Dieter Block. The perceptron: A model for brain functioning. *Reviews of Modern Physics*, 34(1):123, 1962.

[43] Seymour A. Papert. Some mathematical models of learning. In *London Symposium on Information Theory*. Academic Press, New York, 1961.

[44] Marvin Minsky and Seymour A. Papert. *Perceptrons: An Introduction to Computational Geometry*. MIT Press, 2017.

[45] Olvi L. Mangasarian. Linear and nonlinear separation of patterns by linear programming. *Operations Research*, 13(3):444–452, 1965.

[46] M. A. Aizerman, E. M. Braverman, and L. I. Rozonoer. The Robbins-Monro process and the method of potential functions. *Automation and Remote Control*, 26:1882–1885, 1965.

[47] David J. Hand. *Measurement Theory and Practice: The World Through Quantification*. Wiley, 2010.

[48] David J. Hand. *Measurement: A Very Short Introduction*. Oxford University Press, 2016.

[49] Deborah L. Bandalos. *Measurement Theory and Applications for the Social Sciences*. Guilford Publications, 2018.

[50] Lisa Gitelman. *Raw Data Is an Oxymoron*. MIT Press, 2013.

[51] Elaine Angelino, Nicholas Larus-Stone, Daniel Alabi, Margo Seltzer, and Cynthia Rudin. Learning certifiably optimal rule lists for categorical data. *Journal of Machine Learning Research*, 18(234):1–78, 2018.

[52] George Cybenko. Approximation by superpositions of a sigmoidal function. *Mathematics of Control, Signals and Systems*, 2(4):303–314, 1989.

[53] Andrew R. Barron. Universal approximation bounds for superpositions of a sigmoidal function. *Transactions on Information Theory*, 39(3):930–945, 1993.

[54] Gilles Pisier. Remarques sur un résultat non publié de B. Maurey. In *Séminaire d'analyse fonctionnelle*. Ecole Polytechnique Centre de Mathematiques, 1980-1981.

[55] Lee K. Jones. A simple lemma on greedy approximation in Hilbert space and convergence rates for projection pursuit regression and neural network training. *Annals of Statistics*, 20(1):608–613, 1992.

[56] Leo Breiman. Hinging hyperplanes for regression, classification, and function approximation. *Transactions on Information Theory*, 39(3):999–1013, 1993.

[57] Ali Rahimi and Benjamin Recht. Random features for large-scale kernel machines. In *Advances in Neural Information Processing Systems*, 2007.

[58] Ali Rahimi and Benjamin Recht. Weighted sums of random kitchen sinks: Replacing minimization with randomization in learning. In *Advances in Neural Information Processing Systems*, 2008.

[59] Youngmin Cho and Lawrence K. Saul. Kernel methods for deep learning. In *Advances in Neural Information Processing Systems*, 2009.

[60] Alan V. Oppenheim, Alan S. Willsky, and S. Hamid Nawab. *Signals and Systems*. Prentice-Hall International, 1997.

[61] Emmanuel J. Candès and Michael B. Wakin. An introduction to compressive sampling. *IEEE Signal Processing Magazine*, 25(2):21–30, 2008.

[62] Kari Karhunen. Über lineare Methoden in der Wahrscheinlichkeitsrechnung. *Annales Academia Scientiarum Fennica Mathematica, Series A*, (37):1–47, 1947.

[63] Michel Loève. Functions aleatoire de second ordre. *Revue Science*, 84:195–206, 1946.

[64] N. Aronszajn. Theory of reproducing kernels. *Transactions of the American Mathematical Society*, 68(3):337–404, 1950.

[65] Emmanuel Parzen. An approach to time series analysis. *The Annals of Mathematical Statistics*, 32(4):951–989, 1961.

[66] Grace Wahba. *Spline Models for Observational Data*. SIAM, 1990.

[67] Bernhard Schölkopf and Alexander J. Smola. *Learning with Kernels: Support Vector Machines, Regularization, Optimization, and Beyond*. MIT Press, 2002.

[68] John Shawe-Taylor and Nello Cristianini. *Kernel Methods for Pattern Analysis*. Cambridge University Press, 2004.

[69] Allan Pinkus. *N-Widths in Approximation Theory*. Springer, 1985.

[70] Jef Akst. Machine, learning, 1951. *The Scientist*, May 2019.

[71] Ali Rahimi and Benjamin Recht. Uniform approximation of functions with random bases. In *Allerton Conference on Communication, Control, and Computing*, 2008.

[72] Amit Daniely, Roy Frostig, and Yoram Singer. Toward deeper understanding of neural networks: The power of initialization and a dual view on expressivity. In *Advances in Neural Information Processing Systems*, 2016.

[73] Arthur Jacot, Franck Gabriel, and Clément Hongler. Neural tangent kernel: Convergence and generalization in neural networks. In *Advances in Neural Information Processing Systems*, pages 8580–8589, 2018.

[74] Dennis Decoste and Bernhard Schölkopf. Training invariant support vector machines. *Machine Learning*, 46(1-3):161–190, 2002.

[75] Vaishaal Shankar, Alex Fang, Wenshuo Guo, Sara Fridovich-Keil, Jonathan Ragan-Kelley, Ludwig Schmidt, and Benjamin Recht. Neural kernels without tangents. In *International Conference on Machine Learning*, 2020.

[76] Herbert Robbins and Sutton Monro. A stochastic approximation method. *The Annals of Mathematical Statistics*, pages 400–407, 1951.

[77] Mert Gürbüzbalaban, Asu Ozdaglar, and Pablo A Parrilo. Why random reshuffling beats stochastic gradient descent. *Mathematical Programming*, pages 1–36, 2019.

[78] Arkadi Nemirovski, Antoli Juditsky, Guanghui Lan, and Alexander Shapiro. Robust stochastic approximation approach to stochastic programming. *SIAM Journal on Optimization*, 19(4):1574–1609, 2009.

[79] Simon S. Du, Xiyu Zhai, Barnabas Poczos, and Aarti Singh. Gradient descent provably optimizes over-parameterized neural networks. In *International Conference on Learning Representations*, 2019.

[80] Stephen J. Wright and Benjamin Recht. *Optimization for Data Analysis*. Cambridge University Press, 2021.

[81] Bernard Widrow and Marcian E. Hoff. Adaptive switching circuits. In *Institute of Radio Engineers, Western Electronic Show and Convention, Convention Record*, pages 96–104, 1960.

[82] A. Nemirovski and D. Yudin. *Problem Complexity and Method Efficiency in Optimization*. Wiley, 1983.

[83] Shai Shalev-Shwartz, Yoram Singer, and Nathan Srebro. Pegasos: Primal estimated sub-GrAdient SOlver for SVM. In *International Conference on Machine Learning*, 2007.

[84] Yurii Nesterov and Arkadi Nemirovskii. *Interior-Point Polynomial Methods in Convex Programming*. SIAM, 1994.

[85] Tengyuan Liang, Alexander Rakhlin, and Xiyu Zhai. On the multiple descent of minimum-norm interpolants and restricted lower isometry of kernels. In *Conference on Learning Theory*, 2020.

[86] Kaiming He, Xiangyu Zhang, Shaoqing Ren, and Jian Sun. Deep residual learning for image recognition. In *Computer Vision and Pattern Recognition*, 2016.

[87] Yanping Huang, Youlong Cheng, Ankur Bapna, Orhan Firat, Dehao Chen, Mia Chen, HyoukJoong Lee, Jiquan Ngiam, Quoc V Le, Yonghui Wu, and Zhifeng Chen. Gpipe: Efficient training of giant neural networks using pipeline parallelism. *Advances in Neural Information Processing Systems*, 32:103–112, 2019.

[88] Vladimir Vapnik. *Statistical Larning Theory*. Wiley, 1998.

[89] Christopher J. C. Burges. A tutorial on support vector machines for pattern recognition. *Data Mining and Knowledge Discovery*, 2(2):121–167, 1998.

[90] Shai Shalev-Shwartz, Ohad Shamir, Nathan Srebro, and Karthik Sridharan. Learnability, stability and uniform convergence. *Journal of Machine Learning Research*, 11(Oct):2635–2670, 2010.

[91] Olivier Bousquet and André Elisseeff. Stability and generalization. *Journal of Machine Learning Research*, 2(Mar):499–526, 2002.

[92] Shai Shalev-Shwartz and Shai Ben-David. *Understanding Machine Learning: From Theory to Algorithms*. Cambridge University Press, 2014.

[93] Mikhail Belkin, Daniel Hsu, Siyuan Ma, and Soumik Mandal. Reconciling modern machine-learning practice and the classical bias-variance trade-off. *Proceedings of the National Academy of Sciences*, 2019.

[94] Behnam Neyshabur, Ryota Tomioka, and Nathan Srebro. In search of the real inductive bias: On the role of implicit regularization in deep learning. *arXiv:1412.6614*, 2014.

[95] Chiyuan Zhang, Samy Bengio, Moritz Hardt, Benjamin Recht, and Oriol Vinyals. Understanding deep learning requires rethinking generalization. In *International Conference on Learning Representations*, 2017.

[96] Robert E Schapire, Yoav Freund, Peter Bartlett, Wee Sun Lee, et al. Boosting the margin: A new explanation for the effectiveness of voting methods. *The Annals of Statistics*, 26(5):1651–1686, 1998.

[97] Tong Zhang and Bin Yu. Boosting with early stopping: Convergence and consistency. *The Annals of Statistics*, 33:1538–1579, 2005.

[98] Matus Telgarsky. Margins, shrinkage, and boosting. In *International Conference on Machine Learning*, 2013.

[99] Sham M. Kakade, Karthik Sridharan, and Ambuj Tewari. On the complexity of linear prediction: Risk bounds, margin bounds, and regularization. In *Advances in Neural Information Processing Systems*, pages 793–800, 2009.

[100] Peter L. Bartlett and Shahar Mendelson. Rademacher and Gaussian complexities: Risk bounds and structural results. *Journal of Machine Learning Research*, 3(Nov):463–482, 2002.

[101] Vladimir Koltchinskii and Dmitry Panchenko. Empirical margin distributions and bounding the generalization error of combined classifiers. *The Annals of Statistics*, 30(1):1–50, 2002.

[102] Peter L. Bartlett. The sample complexity of pattern classification with neural networks: the size of the weights is more important than the size of the network. *Transactions on Information Theory*, 44(2):525–536, 1998.

[103] Moritz Hardt, Benjamin Recht, and Yoram Singer. Train faster, generalize better: Stability of stochastic gradient descent. In *International Conference on Machine Learning*, 2016.

[104] Tengyuan Liang and Benjamin Recht. Interpolating classifiers make few mistakes. *arXiv:2101.11815*, 2021.

[105] Behnam Neyshabur, Srinadh Bhojanapalli, David McAllester, and Nati Srebro. Exploring generalization in deep learning. In *Advances in Neural Information Processing Systems*, pages 5947–5956, 2017.

[106] Gintare Karolina Dziugaite and Daniel M Roy. Computing nonvacuous generalization bounds for deep (stochastic) neural networks with many more parameters than training data. *arXiv:1703.11008*, 2017.

[107] Sanjeev Arora, Rong Ge, Behnam Neyshabur, and Yi Zhang. Stronger generalization bounds for deep nets via a compression approach. *arXiv:1802.05296*, 2018.

[108] Chiyuan Zhang, Qianli Liao, Alexander Rakhlin, Karthik Sridharan, Brando Miranda, Noah Golowich, and Tomaso Poggio. Theory of deep learning III: Generalization properties of SGD. Technical report, Discussion paper, Center for Brains, Minds and Machines (CBMM). Preprint, 2017.

[109] Moritz Hardt. Generalization in overparameterized models. In Tim Roughgarden, editor, *Beyond the Worst-Case Analysis of Algorithms*, page 486–505. Cambridge University Press, 2021.

[110] David G. Lowe. Distinctive image features from scale-invariant keypoints. *International Journal of Computer Vision*, 60(2):91–110, 2004.

[111] Navneet Dalal and Bill Triggs. Histograms of oriented gradients for human detection. In *Computer Vision and Pattern Recognition*, 2005.

[112] Pedro F. Felzenszwalb, Ross B. Girshick, David McAllester, and Deva Ramanan. Object detection with discriminatively trained part-based models. *IEEE Transactions on Pattern Analysis and Machine Intelligence*, 32(9):1627–1645, 2009.

[113] Peter Auer, Mark Herbster, and Manfred K. Warmuth. Exponentially many local minima for single neurons. In *Advances in Neural Information Processing Systems*, 1996.

[114] Van H. Vu. On the infeasibility of training neural networks with small mean-squared error. *Transactions on Information Theory*, 44(7):2892–2900, 1998.

[115] Surbhi Goel, Adam Klivans, Pasin Manurangsi, and Daniel Reichman. Tight hardness results for training depth-2 ReLU networks. In *Innovations in Theoretical Computer Science*, 2021.

[116] Sergey Ioffe and Christian Szegedy. Batch normalization: Accelerating deep network training by reducing internal covariate shift. In *International Conference on Machine Learning*, pages 448–456. PMLR, 2015.

[117] Yuxin Wu and Kaiming He. Group normalization. In *European Conference on Computer Vision*, pages 3–19, 2018.

[118] Eric B. Baum and David Haussler. What size net gives valid generalization? In *Advances in Neural Information Processing Systems*, 1988.

[119] Peter L. Bartlett. The sample complexity of pattern classification with neural networks: The size of the weights is more important than the size of the network. *IEEE Transactions on Information Theory*, 44(2):525–536, 1998.

[120] Reinhard Heckel and Mahdi Soltanolkotabi. Compressive sensing with untrained neural networks: Gradient descent finds a smooth approximation. In *International Conference on Machine Learning*, pages 4149–4158. PMLR, 2020.

[121] David Page. `https://myrtle.ai/learn/how-to-train-your-resnet/`, 2020.

[122] Andreas Griewank and Andrea Walther. *Evaluating Derivatives: Principles and Techniques of Algorithmic Differentiation*. SIAM, 2nd edition, 2008.

[123] James Bradbury, Roy Frostig, Peter Hawkins, Matthew James Johnson, Chris Leary, Dougal Maclaurin, George Necula, Adam Paszke, Jake VanderPlas, Skye Wanderman-Milne, and Qiao Zhang. JAX: Composable transformations of Python+NumPy programs, 2018.

[124] Alex Krizhevsky, Ilya Sutskever, and Geoffrey Hinton. ImageNet classification with deep convolutional neural networks. In *Advances in Neural Information Processing Systems*, 2012.

[125] Moritz Hardt and Tengyu Ma. Identity matters in deep learning. In *International Conference on Learning Representations*, 2017.

[126] Peter L. Bartlett, Dylan J. Foster, and Matus J. Telgarsky. Spectrally-normalized margin bounds for neural networks. In *Advances in Neural Information Processing Systems*, pages 6240–6249, 2017.

[127] Noah Golowich, Alexander Rakhlin, and Ohad Shamir. Size-independent sample complexity of neural networks. In *Conference on Learning Theory*, pages 297–299, 2018.

[128] Richard O. Duda and Peter E. Hart. *Pattern Classification and Scene Analysis*. Wiley New York, 1973.

[129] Trevor Hastie, Robert Tibshirani, and Jerome Friedman. *The Elements of Statistical Learning: Data Mining, Inference, and Prediction (Corrected 12th printing)*. Springer, 2017.

[130] Xiaochang Li and Mara Mills. Vocal features: From voice identification to speech recognition by machine. *Technology and Culture*, 60(2):S129–S160, 2019.

[131] John S. Garofolo, Lori F. Lamel, William M. Fisher, Jonathan G. Fiscus, and David S. Pallett. DARPA TIMIT acoustic-phonetic continous speech corpus CD-ROM. NIST speech disc 1-1.1. *STIN*, 93:27403, 1993.

[132] Allison Koenecke, Andrew Nam, Emily Lake, Joe Nudell, Minnie Quartey, Zion Mengesha, Connor Toups, John R Rickford, Dan Jurafsky, and Sharad Goel. Racial disparities in automated speech recognition. *Proceedings of the National Academy of Sciences*, 117(14):7684–7689, 2020.

[133] David Aha. Personal communication, 2020.

[134] Frances Ding, Moritz Hardt, John Miller, and Ludwig Schmidt. Retiring adult: New datasets for fair machine learning. In *Advances in Neural Information Processing Systems*, 2021.

[135] Wilbur H. Highleyman and Louis A. Kamentsky. A generalized scanner for pattern- and character-recognition studies. In *Western Joint Computer Conference*, page 291–294, 1959.

[136] Wilbur H. Highleyman. Character recognition system, 1961. US Patent 2,978,675.

[137] Wilbur H. Highleyman and Louis A. Kamentsky. Comments on a character recognition method of Bledsoe and Browning. *IRE Transactions on Electronic Computers*, EC-9(2):263–263, 1960.

[138] Woodrow Wilson Bledsoe. Further results on the n-tuple pattern recognition method. *IRE Transactions on Electronic Computers*, EC-10(1):96–96, 1961.

[139] Chao Kong Chow. A recognition method using neighbor dependence. *IRE Transactions on Electronic Computers*, EC-11(5):683–690, 1962.

[140] Wilbur H. Highleyman. Data for character recognition studies. *IEEE Transactions on Electronic Computers*, EC-12(2):135–136, 1963.

[141] Wilbur H. Highleyman. The design and analysis of pattern recognition experiments. *The Bell System Technical Journal*, 41(2):723–744, 1962.

[142] John H. Munson, Richard O. Duda, and Peter E. Hart. Experiments with Highleyman's data. *IEEE Transactions on Computers*, C-17(4):399–401, 1968.

[143] Yann LeCun, Léon Bottou, Yoshua Bengio, and Patrick Haffner. Gradient-based learning applied to document recognition. *Proceedings of the IEEE*, 86(11):2278–2324, 1998.

[144] Yann LeCun. `http://yann.lecun.com/exdb/mnist/`. Accessed 10-31-2021.

[145] Patrick J. Grother. NIST special database 19 handprinted forms and characters database. *National Institute of Standards and Technology*, 1995.

[146] Chhavi Yadav and Léon Bottou. Cold case: The lost MNIST digits. In *Advances in Neural Information Processing Systems*, pages 13443–13452, 2019.

[147] Jane Bromley and Eduard Sackinger. Neural-network and k-nearest-neighbor classifiers. *Rapport Technique*, pages 11359–910819, 1991.

[148] Jia Deng, Wei Dong, Richard Socher, Li-Jia Li, Kai Li, and Li Fei-Fei. ImageNet: A large-scale hierarchical image database. In *Computer Vision and Pattern Recognition*, pages 248–255, 2009.

[149] Olga Russakovsky, Jia Deng, Hao Su, Jonathan Krause, Sanjeev Satheesh, Sean Ma, Zhiheng Huang, Andrej Karpathy, Aditya Khosla, Michael Bernstein, Alexander C. Berg, and Fei-Fei Li. ImageNet large scale visual recognition challenge. *International Journal of Computer Vision*, 115(3):211–252, 2015.

[150] Jitendra Malik. What led computer vision to deep learning? *Communications of the ACM*, 60(6):82–83, 2017.

[151] Mary L. Gray and Siddharth Suri. *Ghost Work: How to Stop Silicon Valley from Building a New Global Underclass*. Eamon Dolan Books, 2019.

[152] Horia Mania, John Miller, Ludwig Schmidt, Moritz Hardt, and Benjamin Recht. Model similarity mitigates test set overuse. In *Advances in Neural Information Processing Systems*, 2019.

[153] Horia Mania and Suvrit Sra. Why do classifier accuracies show linear trends under distribution shift? *arXiv:2012.15483*, 2020.

[154] Joy Buolamwini and Timnit Gebru. Gender shades: Intersectional accuracy disparities in commercial gender classification. In *Conference on Fairness, Accountability and Transparency*, pages 77–91, 2018.

[155] Tolga Bolukbasi, Kai-Wei Chang, James Y. Zou, Venkatesh Saligrama, and Adam T. Kalai. Man is to computer programmer as woman is to homemaker? Debiasing word embeddings. *Advances in Neural Information Processing Systems*, 29:4349–4357, 2016.

[156] Hila Gonen and Yoav Goldberg. Lipstick on a pig: Debiasing methods cover up systematic gender biases in word embeddings but do not remove them. *arXiv:1903.03862*, 2019.

[157] Arvind Narayanan and Vitaly Shmatikov. Robust de-anonymization of large sparse datasets. In *Symposium on Security and Privacy*, pages 111–125. IEEE, 2008.

[158] Cynthia Dwork, Adam Smith, Thomas Steinke, and Jonathan Ullman. Exposed! A survey of attacks on private data. *Annual Review of Statistics and Its Application*, 4:61–84, 2017.

[159] Cynthia Dwork and Aaron Roth. The algorithmic foundations of differential privacy. *Foundations and Trends in Theoretical Computer Science*, 9(3-4):211–407, 2014.

[160] Amanda Levendowski. How copyright law can fix artificial intelligence's implicit bias problem. *Wash. L. Rev.*, 93:579, 2018.

[161] Robyn M. Dawes, David Faust, and Paul E. Meehl. Clinical versus actuarial judgment. *Science*, 243(4899):1668–1674, 1989.

[162] Ibrahim Chaaban and Michael R. Scheessele. Human performance on the USPS database. Report, Indiana University South Bend, 2007.

[163] Peter Eckersley, Yomna Nasser, et al. EFF AI progress measurement project. `https://eff.org/ai/metrics`, 2017.

[164] Kaiming He, Xiangyu Zhang, Shaoqing Ren, and Jian Sun. Delving deep into rectifiers: Surpassing human-level performance on ImageNet classification. In *International Conference on Computer Vision*, pages 1026–1034, 2015.

[165] Vaishaal Shankar, Rebecca Roelofs, Horia Mania, Alex Fang, Benjamin Recht, and Ludwig Schmidt. Evaluating machine accuracy on ImageNet. In *International Conference on Machine Learning*, 2020.

[166] Benjamin Recht, Rebecca Roelofs, Ludwig Schmidt, and Vaishaal Shankar. Do ImageNet classifiers generalize to ImageNet? In *International Conference on Machine Learning*, pages 5389–5400, 2019.

[167] Timnit Gebru, Jamie Morgenstern, Briana Vecchione, Jennifer Wortman Vaughan, Hanna Wallach, Hal Daumé III, and Kate Crawford. Datasheets for datasets. *arXiv:1803.09010*, 2018.

[168] Eun Seo Jo and Timnit Gebru. Lessons from archives: Strategies for collecting sociocultural data in machine learning. In *Conference on Fairness, Accountability, and Transparency*, pages 306–316, 2020.

[169] Board of Governors of the Federal Reserve System. Report to the congress on credit scoring and its effects on the availability and affordability of credit. `https://www.federalreserve.gov/boarddocs/rptcongress/creditscore/`, 2007.

[170] Nancy E. Reichman, Julien O. Teitler, Irwin Garfinkel, and Sara S. McLanahan. Fragile families: Sample and design. *Children and Youth Services Review*, 23(4-5):303–326, 2001.

[171] Cynthia Dwork, Vitaly Feldman, Moritz Hardt, Toniann Pitassi, Omer Reingold, and Aaron Roth. Preserving statistical validity in adaptive data analysis. In *Symposium on the Theory of Computing*, pages 117–126, 2015.

[172] Cynthia Dwork, Vitaly Feldman, Moritz Hardt, Toniann Pitassi, Omer Reingold, and Aaron Roth. The reusable holdout: Preserving validity in adaptive data analysis. *Science*, 349(6248):636–638, 2015.

[173] David A. Freedman. A note on screening regression equations. *The American Statistician*, 37(2):152–155, 1983.

[174] Avrim Blum and Moritz Hardt. The Ladder: A reliable leaderboard for machine learning competitions. In *International Conference on Machine Learning*, pages 1006–1014, 2015.

[175] Danah Boyd and Kate Crawford. Critical questions for big data: Provocations for a cultural, technological, and scholarly phenomenon. *Information, Communication & Society*, 15(5):662–679, 2012.

[176] Zeynep Tufekci. Big questions for social media big data: Representativeness, validity and other methodological pitfalls. In *AAAI Conference on Weblogs and Social Media*, 2014.

[177] Zeynep Tufekci. Engineering the public: Big data, surveillance and computational politics. *First Monday*, 2014.

[178] Mimi Onuoha. The point of collection. *Data & Society: Points*, 2016.

[179] Amandalynne Paullada, Inioluwa Deborah Raji, Emily M. Bender, Emily Denton, and Alex Hanna. Data and its (dis)contents: A survey of dataset development and use in machine learning research. *arXiv:2012.05345*, 2020.

[180] Alexandra Olteanu, Carlos Castillo, Fernando Diaz, and Emre Kiciman. Social data: Biases, methodological pitfalls, and ethical boundaries. *Frontiers in Big Data*, 2:13, 2019.

[181] Nick Couldry and Ulises A. Mejias. Data colonialism: Rethinking big data's relation to the contemporary subject. *Television & New Media*, 20(4):336–349, 2019.

[182] Peter J. Bickel, Eugene A. Hammel, and J. William O'Connell. Sex bias in graduate admissions: Data from Berkeley. *Science*, 187(4175):398–404, 1975.

[183] Linda L. Humphrey, Benjamin K. S. Chan, and Harold C. Sox. Postmenopausal hormone replacement therapy and the primary prevention of cardiovascular disease. *Annals of Internal Medicine*, 137(4):273–284, 08 2002.

[184] Joseph Berkson. Limitations of the application of fourfold table analysis to hospital data. *International Journal of Epidemiology*, 43(2):511–515, 2014. Reprint.

[185] Judea Pearl. *Causality*. Cambridge University Press, 2009.

[186] Jonas Peters, Dominik Janzing, and Bernhard Schölkopf. *Elements of Causal Inference*. MIT Press, 2017.

[187] Judea Pearl, Madelyn Glymour, and Nicholas P. Jewell. *Causal Inference in Statistics: A Primer*. Wiley, 2016.

[188] Judea Pearl and Dana Mackenzie. *The Book of Why: The New Science of Cause and Effect*. Basic Books, 2018.

[189] Edward H. Simpson. The interpretation of interaction in contingency tables. *Journal of the Royal Statistical Society: Series B (Methodological)*, 13(2):238–241, 1951.

[190] Miguel A. Hernán, David Clayton, and Niels Keiding. The Simpson's paradox unraveled. *International Journal of Epidemiology*, 40(3):780–785, 03 2011.

[191] Peter Spirtes, Clark N. Glymour, Richard Scheines, David Heckerman, Christopher Meek, Gregory Cooper, and Thomas Richardson. *Causation, Prediction, and Search*. MIT Press, 2000.

[192] Bernhard Schölkopf. Causality for machine learning. *arXiv:1911.10500*, 2019.

[193] Jerzy Neyman. Sur les applications de la théorie des probabilités aux experiences agricoles: Essai des principes. *Roczniki Nauk Rolniczych*, 10:1–51, 1923.

[194] Donald B. Rubin. Causal inference using potential outcomes: Design, modeling, decisions. *Journal of the American Statistical Association*, 100(469):322–331, 2005.

[195] Guido W. Imbens and Donald B. Rubin. *Causal Inference for Statistics, Social, and Biomedical Sciences*. Cambridge University Press, 2015.

[196] Joshua D. Angrist and Jörn-Steffen Pischke. *Mostly Harmless Econometrics: An Empiricist's Companion*. Princeton University Press, 2008.

[197] Miguel A. Hernán and James Robins. *Causal Inference: What If*. Boca Raton: Chapman & Hall/CRC, 2020.

[198] Stephen L. Morgan and Christopher Winship. *Counterfactuals and Causal Inference*. Cambridge University Press, 2014.

[199] Angus Deaton and Nancy Cartwright. Understanding and misunderstanding randomized controlled trials. *Social Science & Medicine*, 210:2–21, 2018.

[200] Stefan Wager and Susan Athey. Estimation and inference of heterogeneous treatment effects using random forests. *Journal of the American Statistical Association*, 113(523):1228–1242, 2018.

[201] Douglas Almond, Joseph J. Doyle Jr, Amanda E. Kowalski, and Heidi Williams. Estimating marginal returns to medical care: Evidence from at-risk newborns. *The Quarterly Journal of Economics*, 125(2):591–634, 2010.

[202] Prashant Bharadwaj, Katrine Vellesen Løken, and Christopher Neilson. Early life health interventions and academic achievement. *American Economic Review*, 103(5):1862–91, 2013.

[203] Adriana Camacho and Emily Conover. Manipulation of social program eligibility. *American Economic Journal: Economic Policy*, 3(2):41–65, 2011.

[204] Miguel Urquiola and Eric Verhoogen. Class-size caps, sorting, and the regression-discontinuity design. *American Economic Review*, 99(1):179–215, 2009.

[205] Adam S. Chilton and Marin K. Levy. Challenging the randomness of panel assignment in the federal courts of appeals. *Cornell L. Rev.*, 101:1, 2015.

[206] Kjell Benson and Arthur J. Hartz. A comparison of observational studies and randomized, controlled trials. *New England Journal of Medicine*, 342(25):1878–1886, 2000.

[207] John Concato, Nirav Shah, and Ralph I. Horwitz. Randomized, controlled trials, observational studies, and the hierarchy of research designs. *New England Journal of Medicine*, 342(25):1887–1892, 2000.

[208] David A. Freedman. Statistical models and shoe leather. *Sociological Methodology*, pages 291–313, 1991.

[209] Tim Hwang. *Subprime Attention Crisis*. Farrar, Strauss and Giroux, 2020.

[210] Brett R. Gordon, Florian Zettelmeyer, Neha Bhargava, and Dan Chapsky. A comparison of approaches to advertising measurement: Evidence from big field experiments at Facebook. *Marketing Science*, 38(2):193–225, 2019.

[211] Dean Eckles, Brian Karrer, and Johan Ugander. Design and analysis of experiments in networks: Reducing bias from interference. *Journal of Causal Inference*, 5(1), 2016.

[212] John E. Roemer. *How We Cooperate: A Theory of Kantian Optimization*. Yale University Press, 2019.

[213] Susan Athey and Guido W. Imbens. The state of applied econometrics: Causality and policy evaluation. *Journal of Economic Perspectives*, 31(2):3–32, 2017.

[214] Ioana E. Marinescu, Patrick N. Lawlor, and Konrad P. Kording. Quasi-experimental causality in neuroscience and behavioural research. *Nature Human Behaviour*, 2(12):891–898, 2018.

[215] Dimitri P. Bertsekas. *Dynamic Programming and Optimal Control*, volume 1. Athena Scientific, 4th edition, 2017.

[216] Vincent D. Blondel and John N. Tsitsiklis. A survey of computational complexity results in systems and control. *Automatica*, 36(9):1249–1274, 2000.

[217] Christos H. Papadimitriou and John N. Tsitsiklis. The complexity of Markov Decision Processes. *Mathematics of Operations Research*, 12(3):441–450, 1987.

[218] Benjamin Recht. A tour of reinforcement learning: The view from continuous control. *Annual Review of Control, Robotics, and Autonomous Systems*, 2, 2019.

[219] Francesco Borrelli, Alberto Bemporad, and Manfred Morari. *Predictive Control for Linear and Hybrid Systems*. Cambridge University Press, 2017.

[220] Stephen Boyd. EE363: Linear dynamical systems. Notes available at `https://stanford.edu/class/ee363/`, 2009.

[221] Martin L. Puterman. *Markov Decision Processes: Discrete Stochastic Dynamic Programming*. Wiley-Interscience, 1994.

[222] Herbert A. Simon. Dynamic programming under uncertainty with a quadratic criterion function. *Econometrica*, 24(1):74–81, 1956.

[223] Henri Theil. A note on certainty equivalence in dynamic planning. *Econometrica*, 25(2):346–349, 1957.

[224] Peter Auer and Ronald Ortner. UCB revisited: Improved regret bounds for the stochastic multi-armed bandit problem. *Periodica Mathematica Hungarica*, 61(1-2):55–65, 2010.

[225] Sampath Kannan, Jamie H. Morgenstern, Aaron Roth, Bo Waggoner, and Zhiwei Steven Wu. A smoothed analysis of the greedy algorithm for the linear contextual bandit problem. In *Advances in Neural Information Processing Systems*, 2018.

[226] Patrick Hummel and R. Preston McAfee. Machine learning in an auction environment. *Journal of Machine Learning Research*, 17(1):6915–6951, 2016.

[227] Alberto Bietti, Alekh Agarwal, and John Langford. A contextual bandit bake-off. *arXiv:1802.04064*, 2018.

[228] Christopher J. C. H. Watkins and Peter Dayan. Q-learning. *Machine Learning*, 8(3-4):279–292, 1992.

[229] Gavin Adrian Rummery and Mahesan Niranjan. Online Q-learning using connectionist systems. Technical report, CUED/F-INFENG/TR 166, Cambridge University Engineering Dept., 1994.

[230] Ronald J. Williams. Simple statistical gradient-following algorithms for connectionist reinforcement learning. *Machine Learning*, 8(3-4):229–256, 1992.

[231] Leonard A. Rastrigin. About convergence of random search method in extremal control of multi-parameter systems. *Avtomat. i Telemekh.*, 24(11):1467—1473, 1963.

[232] Yurii Nesterov and Vladimir Spokoiny. Random gradient-free minimization of convex functions. *Foundations of Computational Mathematics*, 17(2):527–566, 2017.

[233] Hans-Georg Beyer and Hans-Paul Schwefel. Evolution Strategies—a comprehensive introduction. *Natural Computing*, 1(1):3–52, 2002.

[234] Hans-Paul Schwefel. *Evolutionsstrategie und numerische Optimierung*. PhD thesis, TU Berlin, 1975.

[235] James C. Spall. Multivariate stochastic approximation using a simultaneous perturbation gradient approximation. *Transactions on Automatic Control*, 37(3):332–341, 1992.

[236] Abraham D. Flaxman, Adam T. Kalai, and H. Brendan McMahan. Online convex optimization in the bandit setting: Gradient descent without a gradient. In *Symposium on Discrete Algorithms*, pages 385–394, 2005.

[237] Alekh Agarwal, Ofer Dekel, and Lin Xiao. Optimal algorithms for online convex optimization with multi-point bandit feedback. In *Conference on Learning Theory*, 2010.

[238] Dimitri P. Bertsekas and John N. Tsitsiklis. *Neuro-Dynamic Programming*. Athena Scientific, 1996.

[239] Horia Mania, Stephen Tu, and Benjamin Recht. Certainty equivalence is efficient for linear quadratic control. In *Advances in Neural Information Processing Systems*, 2019.

[240] Max Simchowitz and Dylan Foster. Naive exploration is optimal for online LQR. In *International Conference on Machine Learning*, 2020.

[241] Alekh Agarwal, Nan Jiang, Sham M. Kakade, and Wen Sun. *Reinforcement Learning: Theory and Algorithms*. 2020. Preprint Available at `rltheorybook.github.io`.

[242] Bernardo Ávila Pires and Csaba Szepesvári. Policy error bounds for model-based reinforcement learning with factored linear models. In *Conference on Learning Theory*, 2016.

[243] Ward Whitt. Approximations of dynamic programs, I. *Mathematics of Operations Research*, 3(3):231–243, 1978.

[244] Satinder P. Singh and Richard C. Yee. An upper bound on the loss from approximate optimal-value functions. *Machine Learning*, 16(3):227–233, 1994.

[245] Dimitri P. Bertsekas. Weighted sup-norm contractions in dynamic programming: A review and some new applications. LIDS Tech Report LIDS-P-2884, Department of Electrical Engineering and Computer Science, Massachusetts Institute Technology, 2012.

[246] Dimitri P. Bertsekas. *Abstract Dynamic Programming*. Athena Scientific, 2nd edition, 2018.

[247] Alekh Agarwal, Sham M. Kakade, and Lin F. Yang. Model-based reinforcement learning with a generative model is minimax optimal. In *Conference on Learning Theory*, 2020.

[248] Dimitri P. Bertsekas. *Dynamic Programming and Optimal Control*, volume 2. Athena Scientific, 4th edition, 2012.

[249] Dimitri P. Bertsekas. *Reinforcement Learning and Optimal Control*. Athena Scientific, 2019.

[250] Tor Lattimore and Csaba Szepesvári. *Bandit Algorithms*. Cambridge University Press, 2020.

[251] Aleksandr Aronovich Feldbaum. Dual control theory. *Avtomatika i Telemekhanika*, 21(9):1240–1249, 1960.

[252] Björn Wittenmark. Adaptive dual control methods: An overview. In *Adaptive Systems in Control and Signal Processing*, pages 67–72. Elsevier, 1995.

[253] Gunter Stein. Respect the unstable. *IEEE Control Systems Magazine*, 23(4):12–25, 2003.

[254] Ursula Franklin. *The Real World of Technology*. House of Anansi, 1999.

[255] David Kendrick. Applications of control theory to macroeconomics. In *Annals of Economic and Social Measurement*, volume 5, pages 171–190. NBER, 1976.

[256] Ernst Friedrich Schumacher. *Small Is Beautiful: A Study of Economics as if People Mattered*. Random House, 2011.

Index